THE ECOLOGY OF A
SUMMER HOUSE

The Ecology
of a
Summer House

VINCENT G. DETHIER

The University of Massachusetts Press

Amherst, 1984

TO LOIS

June 22, 1984

Dear Mr. and Mrs. Leisure,
   (or rather Judge and Mrs. Leisure
   if you please)
Tomorrow is the big day. We take off
from Seattle according to schedule...
hopefully, everyone will converge in time.
   Thank you for so kindly welcoming
me into your home. Have a great
summer discovering your own flora & fauna!
                        Sincerely, Peggy

PREFACE

T HIS selection of vignettes of animal life in our bungalow leaves many animals unmentioned, especially such smaller ones as springtails, pill bugs, and small beetles. My purpose has not been to catalogue all resident life, nor has it been to present complete life histories of the animals I have chosen to describe. It has been to stimulate my readers to an awareness and appreciation of the many and complex manifestations of life in this world and of our participation in that biological wonder. It has been to provide an introduction to some of the exquisite beauty and activities that are so easily overlooked.

In its assemblage of animals our summer home is no different from any other in this part of New England. It differs because it is occupied by a biologist. Were it not a breach of hospitality to do so, I could visit the houses of any of my neighbors and catalogue as many kinds of animals, not necessarily the same species as in our house, but certainly as many kinds. It is all in the looking—knowing what to look for and where to search.

East Bluehill, Maine

1 ∘ THE BUNGALOW

"OMEDAY," said my younger son Paul as he startled me out of a daydream, "you're going to roll off the roof if you insist on sleeping up here." He grinned at me from the top of the ladder.

"I was not sleeping," I retorted indignantly. "I'm up here checking on carpenter ants."

We were both correct—in a manner of speaking. I *had* come up to check on carpenter ants, and I *had* probably dozed or, at worst, dreamed.

Where the fieldstone chimney emerged from the hip roof of the bungalow, there was a comfortable spot for relaxing out of the wind. I sat with my back against the stones and my legs stretched up the slope of the roof. From this vantage I had a quarter-compass view of Blue Hill Bay. Southwestward (whence came most of our summer storms) I could see far down Blue Hill Neck. Directly south lay the end of Blue Hill's Long Island.

Southsoutheasterly I could see past Bartlett and Hardwood islands but was denied the ultimate horizon of the open sea by the distant thin line of Placentia and Swan's islands. To the southeast rose Western Mountain of Mount Desert Island and, due east, Cadillac and Sargent mountains. In the middle distance dull green waves wove a ragged woof against the fog-gray warp of the bay.

From the eminence of the roof I also looked down on the smaller trees. Only the tallest, scattered, centenarian white pines, spired red and white spruces, and scented balsam firs tugged my gaze upward. Abandoning them I looked down through the shallow water at the shore, down on the field and the winding dirt road, down on the birds.

One of the wonders of being on a mountain top is that it is possible to look down on a soaring hawk when one normally looks up at it. The reversed positions are as provoking of wonder as the Looking-Glass world of Alice. The rooftop was my mountain, my Looking Glass, more wonderful than looking down from an airplane, which flies too high and too fast. The roof provided a bird's perspective, solitude, and the quiet to hear the songs of the crickets which wafted faintly upward.

"Have you found where the ants are getting in?" asked Paul.

"Yes, they've gone under the flashing around the chimney and probably are nesting in the two-by-sixes that frame the chimney."

"What are you going to do?"

"Ah, that's the question."

Most of my neighbors played host to carpenter ants at one time or other. Their first reactions were dismay bordering on panic. I well appreciate the reasons. One opens a summer cottage early in June and in the process of cleaning out the winter's dust and the spring's cobwebs discovers neat piles of sawdust in which are intermingled fragments of dismembered ants and other insects. Bits of legs, heads, abdomens, and other anatomical detritus of friend and prey alike are cast willy-nilly on the refuse. Remains of the lame, the halt, and the dead of the colony mingle with the undigested parts of assorted prey. Death ends all discrimination. There is no sentimentality in an ant colony.

As often as we swept up these macabre middens, fresh ones appeared. Small wonder that people get the impression that their houses are being consumed about their ears. Yet the destruction does not in any measure match that of termites. Termites do indeed consume houses. Thanks to a gut full of very special protozoans that

can digest cellulose, the termites literally live in their food. For the carpenter ants the wood is an abode, nothing more. As the galleries are carved in rococo splendor, the sawdust is meticulously removed. Food must be procured from the world outside.

The ants of our colony had such a long trip to make that I wondered why the house offered any special advantage. Obviously the founding queen, who would never have to do the arduous shopping, chose the site with little attention to logistics. The trail led from the roof all the way down to one of the foundation piers, then across about fifteen feet of lawn to a large boulder at the edge of the forest. At this point it fanned out in all directions, around, over, and under obstacles. In some cases the byways actually disappeared underground. The main trail was kept scrupulously clean, there being no litter to obstruct passage or offend the eye.

On another occasion I had idled away the better part of an afternoon watching the traffic on this trail. I had expected to see bits of the house being carted away in one direction and an abundance of provender for the colony flowing in the other. To my surprise nothing much was happening; and other casual checks from day to day confirmed the astonishing aimlessness of ant life.

The density of traffic varied from one to nine ants per minute. Now and again an ant would work for a few minutes clearing obstacles, after which she would hustle down the trail. Sometimes the obstacle was dragged to one side of the track. The trail had become a sunken road where it crossed a three-foot patch of moss. Obstacles such as pine needles that had fallen on this section of road were added to the top so that eventually a roof was formed. During the night, spiders spun their webs over the sunken portions and by morning the roof was bespangled with dew.

Many ants, however, paid no attention to obstacles beyond detouring over, under, or around them. While most kept to the path, some occasionally strayed a few inches afield, apparently exploring, only to return within a few seconds. When I placed a garden slug in the trail, the ants recoiled, detoured around, and made no attempts to remove or capture the creature. A cricket, however, attracted the attention of three or four ants and they eventually started to drag the prize toward the house. Obviously they knew in which direction the nest lay. Generally the ants traveled in a jerky, erratic, stop-and-go manner, always giving the appearance of urgent business and great industry. But although some would return with bark lice and other fragile insects, the majority returned empty-jawed. I was impressed by the enormous amount of random motion. Could it be that all this aimless, scatter-brained expenditure of energy worked as a search pattern? I wonder.

We had a window sill and frame occupied by carpenter ants for at least twenty years. When finally we decided to replace the window with a set of three larger ones, the old frame was removed together with its ant colony. The extent of damage was surprisingly limited considering the long period of occupancy and the perpetual piles of sawdust.

When July's first heat wave stimulates the initial nuptial flight of the year, even the most dedicated entomophile experiences some discomfort. Brief though the flight is, lasting a few days, it has its alarming aspects. From nooks, crevices, crannies, from nowhere and everywhere, huge winged female ants pregnant with eggs congregate on windows and screens. They drop from rafters, hurry across the floor, bang into curtains. This is no invasion of hordes, no amazonian army of thousands. What is lacking in numbers, however, is more than compensated for by the formidable size of these future queens, their aggressive search for new territory, their insatiable drive for empire. There is almost an obscenity in the coupling of their bursting gravidity with haste and aggression. The combination conjures up visions of devastating reproductive potential. This is a different view of maternity than that to which we are accustomed.

There was no doubt about it; I was going to have to do something about the carpenter ants. The project had been put off for too many years.

Loathe to abandon immediately the wonder of solitude of my perch I permitted myself the luxury of a few final moments of dreaming. Then it was that I realized that the house was not just an artifact built by human hands and set in the world of nature for human habitation alone. True, its purpose was to shelter my family and me, but the thought that had struck me so suddenly was that the house was a special ecological niche.

Artificial though it was, the bungalow provided a natural physical environment. It had, for example, a particular temperature which, since this was an unheated building, followed the daily march of the sun and the seasons. On hot August days it warmed as the sun rising over Cadillac Mountain drove out the cold of the night that had been trapped within. Upon awakening, we threw open the doors so that the southwesterly breeze when it roughened the bay would first blow out the cold and later temper the full heat of the day that the uninsulated roof accentuated. On stormy days when the fog and rain wet and beaded the world without, the temperature inside remained more equable. If the storm persisted in insinuating itself indoors through cracks between wall and chimney, under doors, or stealthily through the very walls themselves, a fire of spruce and balsam logs baked out the dampness.

Compared to the outside spaces where, on all except the stillest nights, the wind constantly stirred so that all the scents and temperature and humidity gradients were mixed in never-repeating patterns, the air in the bungalow remained tranquil. With the doors closed, the most momentous stirrings were those occasioned by our own passage, by convection currents where the sun's heat shone through the windows and caused dust motes to dance, or, at night, over the lamps where eddies of heat created small updrafts. Of course, when a fire was roaring, a steady draft, without which the logs would have smoked sullenly, flowed imperceptibly through the room. So, inside there were microclimates characteristic of the bungalow. In wintertime the patterns of microclimates changed but even in change they differed from the outside world.

The exterior of the bungalow was no less a particular environment. Here the wind and humidity were those of the fields and forests, but just as the field differed from the forest and the forest from the exposed rocky shores and abandoned granite quarries, so did the house differ from all outside habitats. The wide two-foot eaves on four sides provided eternal shadow, but not pitch darkness, and subdued wind patterns. The weathered cedar shingles provided unnumbered nooks and crannies as well as surfaces various enough to suit all tastes. The chimney mimicked the rocky shore in its substrate and textural offerings.

That the house offered a varied ecology was amply attested by its fauna and flora. This is what struck me so forcefully as I reclined daydreaming on the roof. There was much more to the house than the carpenter ants and us. According to the deed and tax rolls the house belonged to me who paid the taxes, repaired the roof, painted the trim, and trussed up the cranky, loose stone foundation after each winter's frosts had played their pranks. The town records carried the information that the house was occupied by four persons and one dog (male). I knew better. How anthropocentric we humans are! We four persons, Lois, Jehan, Paul, and I were in fact only the caretakers. The bungalow was inhabited inside and out by legions of beings large and small. Dr. John Doolittle's house in Puddleby-on-the Marsh was nothing compared to ours. Furthermore, all his animals had been invited guests.

Some of the bungalow's occupants were, it was true, merely intruders like the woodland and saltmarsh mosquitoes which came to the doors and windows at night to hunt, which found with uncanny accuracy small tears in the screens, which waited patiently at the doors whence they could rush in whenever anyone entered or left. Whenever that anyone was the dog, Gulliver, a whole coterie of mosquitoes swarmed in with him. They "knew" that inside was a happy hunting ground.

Evenings, as we ate our supper in the kitchen at the table next to the casement window, we could see mosquitoes stretching toward the inside, their proboscides far through the screen. Paul and Jehan, remembering a trick of their grandfather's, would seize a probing proboscis whereupon its owner would pull away leaving the proboscis behind. Unlike lizards which can grow new tails when those appendages are seized by a predator, or the crab which will relinquish a leg to escape certain death, mosquitoes cannot grow new proboscides. Considering the torture they inevitably inflicted on me whenever they did gain entry, I confess that I was sadly lacking in compassion, especially when I observed these same mosquitoes, driven by their thirst for blood, madly persisting in their efforts to get through the screen.

I have called the mosquitoes intruders because they did not abide in the house or even use it as a nursery. They were merely aliens that —having bred in brackish algae-choked pools in the ledges above high-tide level, in placid meanders in the nearby brook, or in the well—came in the night to steal our blood.

In June and July they had to compete with the blackflies which came on the same errand. Stories about the blackflies of Maine are legion. In some parts of the world they are aptly named *Simulium damnosum*. The Maine species is *Prosimulium hirtipes*. Now a new species has been discovered in the same area, *Simulium penobscotensis*, and it too, breeding in the clear-running streams, contributes to our woes. Equally at ease in light or dark it roams the house as freely as if it owns it.

No census of the seekers of blood would be complete without mention of the no-see-ums. Among the smallest of insects, these midges pass easily through screens. Only by closing the doors to the porches or by painting the screens with kerosene or repellent can we discourage their entry. It is astounding that a creature so small as to be almost invisible and with a brain that can be seen only with a powerful microscope can lead such a complex life. Here is a creature that can fly, gram for gram, as well as any bird, can control its yaw, pitch, and roll, can navigate to light, can perceive form, can sense heat, can discriminate among scents, and can find its host with an uncomfortably large measure of success. Here is a creature that hungers and thirsts, that indulges in elaborate sexual behavior, that struggles to escape its enemies. It has a complex respiratory system to provide oxygen to burn the food that fuels its flight muscles. It has a full complement of enzymes, a complete digestive system, a galaxy of sense organs, and all the genetic equipment that we ourselves possess. In short, here is a creature that carries out the same natural functions as do we ourselves, and it is too small to be seen clearly!

These, however, are merely the intruders. The real tenants of our bungalow are natives that live year 'round or are summer visitors like us. Like ourselves, many return year after year, and when age or accident stays their return, their places are taken by younger generations that carry on the tradition.

No, our bungalow is not just the abode of us (and the carpenter ants). Nor is it the impoverished world of the urban house which can boast at best of flies, cockroaches, silverfish, and, perhaps, fleas, bedbugs, house mice, and rats. Ours, as I realized that morning on the roof, is a richly populated ecosystem.

2 ° CHIMNEY DWELLERS

$\mathbb{A}$ CHIMNEY is hardly the most promising site at which to begin an ecological study. In Europe where chimneys have been associated with storks in legend and in fact for centuries, the situation would have been different. Here in Maine the only birds that have set up housekeeping in chimneys are swifts. On several occasions in years past swifts had glued their fragile twig and saliva cups to the inside of our chimney, thereby causing us to delay use of the fireplace until swiftian family affairs had been completed. Ever since I had had to cap the chimney with wire screen to arrest sparks, we had not been favored with the company of these birds. We missed their aerial acrobatics and companionable twittering although we still were visited occasionally by other swifts from the village at the head of the cove.

The chimney appeared to be a barren place. To be sure, the rough stone was softened here and there with lichens. Rosettes of a green-gray species that grew commonly on the trunks of the larger surrounding trees decorated all but the seaward side. They shared space with a dun-colored species that also formed rosettes. Because these hugged the rock more closely, were more modestly pigmented, and tended to coalesce into large irregular blotches, they were easily mistaken by the casual eye for mineral inclusions of the rock itself. Also, they were outshone by a striking ochre species. The presence of the ochre lichens was rather puzzling because these occurred neither on the neighboring trees nor on the rocks of the shore. They were common only on the granite ledges of Darling Island which lay a mile offshore.

My attention to these primitive manifestations of life, the only life thus far seen that seemed to find this part of the house congenial, was fleetingly distracted by the thought that this might be a good time to clean the chimney. At irregular intervals we reluctantly attended to this chore. While our technique surely would not have earned the approbation of genuine chimney sweeps, it was both simple and effective. One person on the roof would lower a rope until the end reached the fireplace. At that point whoever was helping would tie on a bundle of freshly cut birch switches. His partner on the roof would then pull up this improvised brush. Several repetitions of the maneuver with a fresh bundle each time did the job. Paul, however, had judiciously disappeared from sight, and Jehan, judging from the sound of hammering, was somewhere on the beach. Besides, it was more entertaining to examine lichens. Until this moment I had never paid much attention to them beyond admiring the aesthetic contribution that they made to the deep woods and the shoreline boulders above the high-tide mark. So I stirred myself to the point of going down into the house to get a hand lens.

A lichen is one of the abiding curiosities of the biological world. It is not a living organism; it is two living organisms, an alga and a fungus living in such an intimate mutualistic partnership that together they present to the world a unity whose characteristics transcend those of either partner alone. This unity and characteristic identity of each "species" of lichen poses a continuous problem to their classifiers because the alga is a true species entitled to its own name, as is also the fungus. When either is grown in culture, it breeds true to its species. Together in culture each goes its own way. Only in nature do they cooperate to form a lichen, the alga providing nutrients for both by photosynthesis and the fungus providing protection, holding moisture, and probably aiding its own growth by absorbing dissolved chemicals from the rock.

Viewed through a hand lens the green-gray lichen on the chimney resembled patterns ranging from one that might have been constructed from an arrangement of miniature oak leaves to some of the foliate frost-crystal patterns that formed on window panes in the wintertime. Scattered on the surface were small shallow cups, the reproductive structures. The ochre lichen was similar in general appearance but possessed stalked cups. The gray lichen, ranging in color from charcoal to silvery gray, resembled more than anything else a nodular lava flow. Many of the nodules were craters filled to the rim with a granular black matrix. The nodules together with the craters gave the appearance at this magnification of a lunar landscape.

Lichens are among the slowest growing plants, and the longest lived. Knowing the rate at which they grow, students of glaciers have sometimes employed lichens to date moraines and thus determine the rate at which a glacier is retreating. By determining the rate at which a lichen rosette increases its diameter one has a yardstick of growth.

Thinking about this I began to wonder whether the lichens on the chimney existed on the rocks before the chimney was built or arrived after the rocks had been cemented into place. There was a way to answer the question. All I had to do was measure the circumference of selected lichens, wait one year, and remeasure them. This would give me the rate of growth, and, making the assumption that the rate was approximately the same each year and knowing the age of the bungalow, I would resolve the question.

The following summer I did have my answer: the gray lichens were one hundred and ten years old; the yellow lichens were thirty years old. The house was seventy years old.

Somehow these curious plants, because they did not move and were, in our gross perspective, two dimensional, belied the qualification of life. Nevertheless, though they did not require our chimney for their existence, they were part of the living ensemble of our bungalow and deserving of attention.

My thoughts regarding the population of the bungalow were reinforced by a trim, black, medium-sized wasp that persisted in buzzing around my head. Although in the usual human way I might say that the wasp was annoying me, my presence at the chimney clearly disturbed her. She continued to explore uncertainly the general region of my head.

There was nothing aggressive about her actions. On the contrary, she seemed anxious—if a wasp could feel anxious. Put more objectively, she had noted something—me—in her environment that had not been there previously. When I changed my position a few feet, she suddenly alighted on the one side of the chimney that was free of lichens. No sooner had she landed, however, than she disappeared from sight. Now it is not unusual for an insect to be so superbly camouflaged that it blends perfectly with its background as soon as it ceases all movement. This wasp, however, had not disappeared into its background, it had disappeared into solid granite. The mystery of the disappearance was compounded a minute later by a sudden reappearance and hasty departure. I searched with the hand lens what I presumed to be the correct area, but discovered nothing. Three minutes later the wasp returned. This time, having been alerted, I watched her most carefully. The mystery was solved. She had vanished into a small, round, wasp-sized hole.

To appreciate the exquisite happenstance of a hole in the rock of just the right size for this small wasp it is necessary to understand something of the composition and weathering characteristics of the local granite from which the chimney was constructed. This is a coarse, light-colored granite composed of feldspar, quartz, and hornblende. A century ago it was quarried extensively for use in buildings and bridges in Boston and New York. When this hard rock is exposed to weathering, the feldspar is the first mineral to be decomposed. Where the granite is exposed to the sea, this action is hastened. As a consequence the feldspar between the white quartz and the black hornblende dissolves leaving miniature caves the sizes of which are determined by the texture of the granite. It so happened that the crystals in this rock were wasp sized. It was obvious, furthermore, from the pattern of weathering, that some of the blocks in the chimney had been quarried from ledges along the shore.

This wasp and others of her tribe had discovered these convenient caves and, just as troglodytes had populated cliffs of soft stone in ancient times, these animals had populated our chimney. I could observe the entire process of cave dwelling here on the roof because no two wasps had reached the same stage of housekeeping.

After exploring a wide range of holes, a wasp would select one. What subtle differences existed between one hole and the next of the same size only she knew. In any event, the chosen hole seemed to require some alteration. The first step involved smoothing and shaping the entrance. This she did by framing the doorway with mud which she molded into a smooth round lip. Its outside circumference was blended into the surrounding rock with a thin collar of mud of rougher texture than the frame.

Where, I wondered, did she find wet clay hereabouts, especially since there had been no rain for two weeks. Old records of drought had fallen this summer. We had been experiencing drought so severe that the woods were closed and all outdoor fires were prohibited. The forests were tinder dry. The ranger in the fire tower on Blue Hill was especially alert for the mere suspicion of smoke, and there was talk of banning the fireworks display at the Fair in September if we did not get some good soaking rains by August.

There was one way to solve the mystery of wet clay, so, descending to the ground, I got my field glasses, a small bottle of opaque water-soluble paint, and a fine camel's-hair brush. Once back on the roof I stationed myself at a point on the ridgepole where I could both see the chimney and look toward the direction in which the wasp departed after each bout of masonry.

When the wasp returned and while all her energies were engaged in plastering, I cautiously and gently dabbed a minute spot of paint on her thorax just between the bases of the wings. Not in the least disturbed, she finished her work, sprang into the air, and flew off in the direction of the bank marking the end of the lawn and the beginning of the beach. At that I rushed back to my perch on the ridgepole. On this trip she was too quick for me, but on the next I was able to follow her with the field glasses and mark the spot between two poplar saplings where she literally dropped out of sight. At that point I needed an assistant.

From the sound of hammering in the direction of the beach, I knew that Jehan still busied himself with the boat.

"Jehan," I called during a lull in the hammering, "will you help me for a few minutes?"

"Where are you?"

"On the roof."

"What are you doing?"

"Watching wasps."

"I thought you were taking care of the carpenter ants," he replied in a tone that carried a hint of admonition.

"All I need you for is to keep your eye on the edge of the bank where the devil's paintbrush is growing between the two poplars. When I shout, a small wasp will appear. I want to know where she goes."

This bizarre request, together with the implication that a shout from me would cause the appearance of a small wasp, was greeted with a significantly long silence. Finally, apparently having decided that if he humored me I would disturb him no longer, he said, "There are bugs all over the place here."

"Well," said I, "this one has a spot of yellow paint on her thorax, er, back."

"O.K."

During this exchange the wasp had returned to work. As soon as she had finished and taken flight, I tracked her with the field glasses.

"Here she comes," I shouted.

The answering shout announced failure. "I didn't see her."

But five flights and thirty minutes later Jehan got her range and located her destination.

By the time I arrived on the beach she had taken off again; however, I could now sit down and wait.

The place to which she had returned was a little oasis in an otherwise sun-baked environment. The bank fell precipitously from lawn to beach and ledge. The heavily browed top sported a wild mixture of devil's paintbrush, dandelions, Queen Anne's lace, feral heliotrope, primroses, and mixed grasses. The face, a favorite nesting place for kingfishers before the bank had become too frowningly browed, showed fine layers of varved clays deposited by some ancient glacial stream. At the foot of the bank where chunks of turf had fallen, a narrow garden bloomed. Here morning-glories, evening primroses, lamb's-quarters, jewelweed, and thistles competed for space. Here also silver-spotted fritillaries busied themselves, along with bumblebees, in the thistle blossoms. It was just at the juncture of the clay cliff and this lower garden that the wasp mined the clay. Here others of her kind shared the mine. The clay, however, was dry and crumbly, hardly the stuff to please the heart of a potter.

A moment later I discovered the secret of potting in dry weather. From a long crack in a seamed granite ledge, protected from all but the highest tides in southwesterly storms, there flowed a miniscule trickle of icy ground water. In fact, the erosion of the bank owed less to the sea than to inland ground water.

In modest manner the insignificant trickle resembled a water hole in a dry savanna. It lacked the spectacular clientele of African water holes. In place of zebras, wildebeests, impalas, and others, it attracted bees, flowerflies, butterflies, and an assortment of wasps, species that no other circumstance than need for water would have prompted. There were midges, a tiger swallowtail, the wasp from the chimney, and a half-dozen others of her species. Some of the insects were drawn by thirst, some by a watery urge to oviposit, some, like the tiger swallowtail, to suck up whatever dissolved salts the trickle released as it flowed down over rocks in the intertidal zone. The honeybees came to collect water to cool the hive, and the wasps, to gather water for mud pies.

The marked wasp drank until it appeared she would burst. Then, water-sated, she flew to the little talus slope at the base of the clay cliff. Here she regurgitated the water. With her mandibles and palps she worked the wetted clay to the consistency of beaten cream, formed it into a round pellet, and flew away in the direction of the bungalow.

This was to be her last mud-gathering trip for a while. The doorway was completed to her satisfaction. This is to say, whatever stimulus the hole in the rock had provided for masonry had now been terminated by the execution of that work. The wasp's behavior entered another phase, the egg-laying phase. She entered the cave where she busied herself for two or three minutes with laying a single egg. No sooner had this task been completed than her behavior switched to the hunting phase. After she had departed I was able with the hand lens to discern faintly in the darkness a thin ovoid egg glued at one end to the roof.

The practical sequencing of behavior is one of the more interesting aspects of the activity of animals. Creatures that do not appear to reason are programmed, as it were, to execute chains of behavior each link of which is initiated by some particularly significant stimulus, is carried to completion, and by its consummation eliminates that stimulus and provides for another to replace it in significance.

Having laid her egg, the wasp now was intent upon hunting. For me to follow her on these forays was manifestly impossible. Instead I settled down at the chimney to record the results. In the course of the day, and naturally I only checked periodically, she packed the cave with eight caterpillars each about as long as she was. These were leaf-roller caterpillars (Tortricidae) which she had, through diligent searching, found on some tree and extricated from their curled leaf houses. All were of the same species. Each had been paralyzed with a sting when captured, thoroughly kneaded by repeated biting, and brought back to be packed carefully into the cave.

Provisioning continued into the next day. Presumably the amount of time required depended on the distance from nest to hunting ground and upon the abundance of prey. If a rainy or foggy day intervened, all activity ceased until the sun shone again, but in this particular instance the weather held clear. Each day the sun made its scheduled appearance over Cadillac Mountain, the bay rippled only to the dive of a cormorant and the surfacing of a harbor seal. The wind did not arise until after nine o'clock, nor did the wasp. Wherever she had spent the night, it took a warmer sun than a newly risen one to stir her to activity.

On her first trip of the day, the third day, to the nest she spent many minutes inspecting the site and its immediate vicinity. Four trips to the water hole and clay bank were required to close the cave. The ball of mud procured on the first trip, which took five minutes, was spread around the lip of the entrance and worked carefully toward the center. It sufficed to close all but a small central hole. The wasp labored at the task for one minute. The second trip required three minutes. With the second ball of mud, the wasp completed the cap over the doorway. The third trip, also three minutes long, supplied mud to reinforce the cap and to begin blending it to the surrounding rock. The fourth and last trip, two minutes long, provided mud for putting finishing touches on camouflaging the entrance. Mud from the previous trip had served quite well in this regard because not only did I have trouble picking out the nest from the texture of the granite but so also did the wasp. At least, upon returning with the last mud ball she landed on the wrong spot. I thought for a moment that the mud would dry before she did locate the nest but she found it in time and used the mud to rectify some imperfections that her eye or sensitive tapping palps had perceived.

That task completed, the wasp paused to clean her antennae. She then departed, never to see her offspring. Presumably all memory of that nest, its location, what stage of provisioning or construction it was in, and how to navigate to it from a distance had been erased. The egg, incubated by the sun-warmed rock, would give rise to a mealy-looking grub which would find itself in a larder provisioned with live but harmlessly twitching caterpillars. Wasp venom took the place of refrigeration, and all was well with the grub which would pass the winter in the cave in a cocoon of its own spinning and emerge the following summer as a wasp—unless!

A significant question remained. One might imagine that superbly camouflaged caves in solid granite were secure, impregnable. And one is reminded of a genre of detective story, the locked-room type, in which the murderer could not possibly gain entrance, kill, and leave the corpse in the room still locked from the inside. Sometimes the grub in the cave was murdered, and, as in the detective story, the murderer had to be in the room before the door was locked. Occasionally a truly minute parasitic wasp manages to sneak into the cave before it is sealed and to lay her own egg which will produce its grub developing in the egg of the mud wasp. Camouflage and impregnability are of no avail. The parasite does not have to find the cave. She merely shadows the tenant, follows her home, and enters the unguarded cave after the owner has departed.

All this activity, adventure, plots, and counterplots go on in our chimney. Some wasps even take advantage of drill holes when erosion holes do not suit. In some of the granite blocks that were not dressed with care a few shallow drill holes remained. Because of their large size they did not accommodate wasps as snugly as dissolved caves, but the lack of precise fit was compensated for by a large amount of mud stuffing.

Every summer this cycle of life goes on in the chimney while all the time the lichens creep over the granite surface with glacial slowness and inexorability. And if this most inhospitable and exposed part of the bungalow caters to such a population, it is hardly an exaggeration to suggest that the house is inhabited literally from chimney to cellar.

For three days a southwesterly storm had kept the whole family indoors. From time to time cabin fever drove us outside where we stood in our oilskins staring out to sea, gauging the pitching of the sloop at her mooring, or tramping to the woodshed for logs to replenish the fire. Seagulls also were abroad as were doubtless other creatures, especially back in the forests where the wind was tamed and the rain only sifted to earth instead of falling in a steady downpour. For the most part we stayed inside.

A huge fire was kept going, more to drive out the dampness than to alleviate the cold; but there was some chill too. At night the bungalow grew cold enough to make the fire welcome. The fireplace accommodated four-foot logs and because of its size radiated warmth through most of the room. It was customary to have a fire always laid, regardless of the weather; that is to say, the kindling and pine and spruce logs simply awaited a lighted match.

For many evenings before the present storm broke I had listened to the creak, creak, creak of a woodsawyer tunneling in one of the logs. As I knew from having exposed some of these larvae while splitting logs, this one was of the cerambycid family. It was a whitish, soft-bodied grub with heavily armored brown jaws and a head full of powerful jaw muscles. Under happier circumstances the grub would have pursued its secretive ways until the day of metamorphosis when it would have emerged as a long-horned beetle, a creature with magnificent curving antennae three times the length of its body.

Sometimes it took these insects twenty or more years to grow to adulthood. If in the meantime the log in which they lived was fashioned into a piece of furniture, a process that bothered them not one whit, the householder was subsequently treated to the spectacle of an enormous beetle emerging from the chest of drawers, desk, chair, or bed.

I had thought of rescuing our longhorn from the log, but it would have been a useless gesture. There would be no placing it successfully in a foster home. Thus, the moment eventually arrived when I lighted the fire on the first day of the storm and sealed the grub's fate. I consoled myself with the thought that it would fall gently into a stupor from the smoke long before the heat reached into its tunnel. Martyrs burned at the stake in times past had received less compassionate thought.

That was two days ago. No longer did a creaking come from the fireplace. Instead there were snaps and crackles and occasionally miniature explosions as bubbles of pitch erupted and flung red embers onto the hearthstone. Since the hearthstone was a wide expanse of granite, only rarely did an ember arch beyond it. On one memorable night years ago a spark had arched high, with humorous results. The butt of the humor was an uncle. The old gentleman used to sit by the fire in a tall wingback chair where he sipped brandy and composed in his mind the letter that he would write to one of his many friends later in the evening. That evening he had left the chair to replenish his glass. Returning he had resumed his reverie but within a few minutes interrupted his thoughts to remark that he was getting uncomfortably hot. The private haze of smoke that enshrouded him told the whole story. While he had been refilling his glass one of the logs had fired an ember into the wingback chair to be incubated by my uncle.

No mishaps occurred this evening, however. The fire had settled down to a gentle hissing and purring, the rain beat steadily on the roof, occasionally whipped to a frenzy by gusts, and all was cozy within—cozy for us and for all the creatures in the house. It was on this evening that the wood mice kept us busy.

For as long as I can remember there had been wood mice in the bungalow. They were year-round residents, true natives. Each summer when we opened the house there would be numerous signs of their winter occupancy despite all efforts to discourage it. Our fall precautions of storing everything that might tempt them (pillows, blankets, clothing), mothballing the rugs and furniture, and wrapping overstuffed furniture in huge plastic bags greatly reduced the damage—but never eliminated it completely. That was as hopeless as trying to achieve a perfect vacuum. Even while we were in residence the mice enjoyed a free run of the house.

Our attitude toward them was painfully ambivalent. Unlike common house mice, these dainty creatures appealed to one's tender instincts. Their white-stockinged feet gave them an air of delicate charm. Their large eyes, proportionally larger than those of other mice, drew immediate attention. In human expression large eyes have always been equated with wonderment, innocence, and helplessness. Their very large ears accentuated the impression of helplessness and vulnerability. I was reminded that the wee people, the leprechauns, dwarfs, and gnomes, are often pictured with disproportionately large ears.

One assumes that large eyes are correlated with nocturnal modes of life. Presumably they are indicative of an enhanced ability to see in dim light, and indeed, experiments with wood mice have shown that they do very well in this respect; however, there are other nocturnal mice that manage very well with eyes of average size.

Insofar as ears are concerned, a large pinna, the fleshy part of the ear, obviously aids in localizing and funneling sound somewhat on the principle of the ear trumpets employed several generations ago by people who were hard of hearing.

It was difficult, as I have said, to reconcile the gentle appearance of the mice with their irresponsible behavior in the bungalow. Observed close at hand they were delightful. From a distance they could only be regarded as destructive.

Until quite recently there was a long wrought-iron curtain rod extending the entire width of the living room, a rod from which hung green burlap drapes. On the particular evening of which I speak (the rod not having yet been removed), our attention was drawn to a mouse that scurried with nimble feet from one end to the other—or was it a traffic of several mice? Having earlier that summer had some choice linen doilies transformed into nesting material neither Lois nor I was in a particularly forgiving mood. As a consequence my objections were feeble indeed when Jehan brought out the BB gun from its resting place behind the door.

"What are you up to?" I inquired.

"I'm going to get the mouse."

"In the house?"

Before I could object the mouse reappeared, and Jehan took a shot at it. The BB buried itself somewhere in the dark rafters near the ridgepole. I felt sure that the mouse was perfectly safe but I was less sanguine about ourselves given the possibility of BB's ricocheting among rafters. I called a halt to that sport.

My stern but belated resolve suffered a setback in the next act of the unfolding drama. I had been stretched out on the sofa near the fireplace, and I soon had a feeling that there was some unusual activity in the sofa. Somewhere in the depth of the back, more particularly next to my head, the upholstery beat like an unsteady heart to the accompaniment of slushes and flutters like leaky valves, and occasional squeaks unlike the sounds of any heart. Compassion was one thing; common sense another. But it was bedtime and the hour was too late to begin dismantling the sofa. In the few hours remaining till morning little damage could be done.

I was not to be let off that easily. Lying in bed listening to night noises I heard rustling in a blanket chest. That was too much. I got up, switched on the light, and began to rummage gingerly among the blankets in the chest. As I turned back a corner of the bottommost blanket, I uncovered a family of wood mice.

Looking down at the mother and four, or was it five, hairless, pink, young mice, each not much larger than a June bug, I was torn between mercy and practicality. In all probability I was going to have to face the same dilemma in the morning when I tackled the sofa. The mice, like all other creatures in the house, lived in the territory of nature's greatest predator, ignorant of the power of life and death that he held over them. Nothing is too large to be slain or too small to be obliterated. In the world at large we can kill them all, from the sperm whale to the virus. What was I to do with the mice?

Although the transgressions of the mice resulted in destruction of our possessions, I knew that there was neither malice nor premeditation in their actions. It is difficult to punish what is natural with death. "But," I admonished myself, "this is sentimental rubbish." And I was reminded of another similar situation in which two of my neighbors had found a small bird which they brought home and attempted to feed. After spending an entire day in fruitless attempts to cram something down its gullet, they telephoned their handyman, a man wise in the ways of the country and, after explaining the situation, they sought his advice.

"What shall we do with this helpless little bird?"

His laconic reply, "Kill the darn thing!"

I could not bring myself to do this to the mice, not because they were "cute" and helpless but because as a biologist I was acutely aware that we can so easily destroy life and are still unable to create it. I went to the kitchen to find a container to house the family till morning.

When I returned, the mother was no longer there. Here was a dilemma. The young, judging from their appearance, were less than ten days old. Their eyes had not yet opened; their toes were still fused together weblike; they were not able to stand and could only tumble about. Clearly, they had not been weaned, so that to transfer them to a new home without their mother was to sentence them to death.

My pondering was interrupted by a call from one of the boys on the porch. This summer they had elected to sleep on the hammock and glider on the screened side porch.

"Hey! Hurry out here; there's a mouse with a baby in her mouth!"

I hurried out.

Paul had the beam of his flashlight trained on the eight-by-eight plate on which the rafters rested. Caught in the light but much too busy to try to avoid it was the mouse. She carried a pink young one by the nape of its neck and, as we watched, scurried along the plate until a rafter blocked her progress. At this point she climed laboriously around the underside of the rafter, then back onto the plate. She had twelve rafters to negotiate before reaching a corner. In the corner, I discovered that some time earlier another nest had been built, mostly with shreds of bark torn from our cedar posts. This last observation did little to deepen my love of these small creatures.

As we watched, the mother arrived at the nest, dumped the pup in, rather unceremoniously I thought, and began the return journey. Always hugging a wall, she traversed the front of the porch to the corner, continued along the side, descended to the floor, passed through the door, ran along the wall of the living room, behind the piano, into the bedroom, along the wall behind the blanket chest, and into it by way of a crack in the back. She retrieved another pup from the nest and began a second arduous journey. I could now see that the litter numbered five.

An intriguing question came to mind. Did she know how many young she had? That is, did she count the number deposited in the nest on the porch and say to herself, so to speak, "Yes, they are all here now"? Apparently not, because after having retrieved all the pups she made one final trip to the nest in the chest. After poking around for about a minute she departed. At this point it was well after midnight; the bungalow was cold; it was time to return to bed. There would be plenty of time tomorrow to deal further with the mice.

The following day the boys and I built a cage from hardware cloth, attached a bottle with a drinking tube, and placed inside a running wheel that we happened to have on hand. Jehan collected the mouse family, nest and all, and established them in their new home. The mother's first action was to explore the new neighborhood. Very thorough she was too.

For the next four weeks we had a first-hand education in bringing up wood mouse families. Among other things I settled the question of the counting of offspring by conducting some simple behavioral experiments. These consisted of removing one or more pups from the nest, which was conveniently located in a corner of the cage, and placing them in the shredded newspaper litter in a far corner. Each time I did this the mother would retrieve as many young as there were; that is, she never left anyone behind, but she did continue to search briefly after everyone was home.

But how did she know that everyone was home or that no more were stranded at the far end of the cage? A clue to her behavior was provided by a second experiment in which three pups were removed to the far corner, two of these were concealed beneath the paper, and the third was isolated in a closed jelly tumbler in plain sight. The mother easily found the two hidden pups and returned them to the nest. All the while she ignored the pup in the bottle. She did not return seeking it. This observation suggested that she was not counting and that she was not locating the lost ones by sight. That left two possibilities: she located either by smell or by sound—or by both combined. A direct way of determining which sense played a dominant role would have been to anesthetize a displaced pup. At this point I really could not bring myself to do the experiment.

My reluctance was rewarded by a cooperative pup that refused to squeak while his two displaced siblings did. Although it is well known, and we had observed, that displaced young give a characteristic cry of abandonment, which is audible to human beings, it is not known whether the cry has an ultrasonic component audible only to mice. Fortunately for inquisitive human observers all cries are accompanied by a rhythmic pulsing of the body so that if a pup is motionless there is a good chance that it is not making a sound. This was the condition of one of the three pups.

The mother, presumably alerted by cries from the far corner of the cage, hurried to that spot. Quickly nuzzling a squeaking pup she picked it up by the nape of its neck and returned it to the nest. Back again she came. The pup that was squeaking lay on its side. She picked him up by the back. She returned a third time. She nuzzled the place where the two had been, nuzzled the silent pup, explored briefly, and then returned to the nest leaving him behind. How long he would have remained lost had I not done the retrieving I cannot say.

During this period I had evicted the other family of mice from the back of the sofa. That family consisted of three nearly mature young

that had not left home. I placed them in a pail with some shredded paper. My intention was to release them somewhere in the woods. Something delayed my doing this, and in the meantime Jehan had begun feeding them. The pail covered with a piece of screen was on a bench on the back porch. The screen was to keep predators out as much as to keep the mice in; nevertheless, they did escape one night. To our great surprise they returned a day later. Either they found the food in the pail more plentiful than in the hard, cruel world outside, or they decided that the pail was as good a home as any.

The question of exactly which of two local species of wood mice our guests were was really one for an expert. A convenient way of distinguishing the two, as a specialist friend of mine pointed out, is by examining the tail. The deer mouse, the one that had nested in the sofa, has a tail on which there are two sharply divided hair colors, black above and white below. There is also a "pencil" of hairs at the top. On the tail of the white-footed mouse the demarcation of black and white hairs is less sharp. Both mice have white-stockinged feet. Most of the mice nesting in the bungalow were deer mice; however, toward the end of summer they had to share space with white-footed mice that eventually spent the winter there.

From time to time during the course of the summer we observed other mice in the bungalow. They seemed to enjoy the crevices of the rough fieldstone chimney and the beams and rafters overhead. One night one even perched on the antlers of the stuffed deer over the fireplace and regarded us without fear. I knew that with the onset of winter more would take up residence. The life span of the average wood mouse in the wild is about two years. Our annual visitors were, therefore, newcomers. The bungalow obviously was a select neighborhood from a wood mouse's point of view. In the long run, as long as we took adequate precautions against their winter mischief, the pleasure that they accorded us of an evening outweighed the damage. To be perfectly honest, there was no way I could have made the bungalow mouse-proof even had I so desired.

My ambivalent attitude toward the mice and their destructive (from our point of view) style of life left me pondering such questions as why it is psychologically easier to exterminate carpenter ants than wood mice. The difference in attitude certainly has something to do with our perceptions of animals and even with how we see ourselves.

Generally speaking, humans tend to empathize with warm, furry, cuddly creatures that look one in the eye as though about to speak. As one compares animals at lower levels on the evolutionary scale, however, the feeling of relationship rapidly attenuates. Birds still evoke a measure of compassion. Again it may be their bright, inquisitive eyes. Those near relatives of birds—snakes, lizards, and turtles—are loved by fewer people. Toads and frogs cannot compete with mammals for our affection despite the fact that the eye of a toad is so beautiful that Shakespeare in *As You Like It* referred to it as a jewel: "the toad, ugly and venomous, wears yet a jewel in his head."

By the time we lower our sight to insects, all sympathy has vanished. These small creatures with their huge, immobile, staring eyes and their bizarre body structure are too alien to our chauvinistic views of acceptable form to evoke any sense of companionship.

The fact remains that, for whatever reason, the wood mice in the bungalow do add a dimension of companionship. The place would not be quite the same without them.

4 ∘ A CEDAR-POST COLONY

WHEN the bungalow was built around the turn of the century, hip roofs, wide eaves, and generous porches on three sides were in style. At the time, northern white cedar grew in abundance almost everywhere. It was natural, therefore, that cedars growing on the site should have been cut to support the porches. The bark had been left intact and now, seventy years later, it is still attached except on a single column of an open porch where one day we surprised a red squirrel busily shredding off long strips for nesting material. She was probably not the only vandal.

I had never favored the cedar posts with more than passing glances until a particular afternoon in July. This day I was sitting on the back porch facing west and in the direction of the flower garden. The slanting sunlight picked out the flying insects and revealed the great density of aerial traffic. The atmosphere, which usually appeared so empty except for the more obvious giants—butterflies, bees, and wasps—was crowded with aphids, midges, small wasps, minute beetles, leafhoppers, and species too numerous to mention. Had I been more interested in cataloguing this assemblage, I could have walked out to the driveway where Paul was painting the 1930

Model A Ford that he had restored. He was at his wit's end trying to keep the air-borne flotsam off the freshly painted surface.

In the garden itself, white and purple Japanese iris were in full bloom as were columbine, marguerita, foxglove, and petunia, visual delights to us, gastronomic delights to the suckers of nectar. I alternately watched a ruby-throated hummingbird at the petunias and foxglove and a group of red-bodied dragonflies cavorting over the lawn. The ruby-throat soon sucked the flowers dry for this day, buzzed over the flower boxes where he apparently found the geraniums unrewarding, and finally shot up to disappear over the tree tops.

The actions of the dragonflies were puzzling, but that was only because I had much to learn about them. These were not hunting. The flight of hawking dragonflies is unmistakable. Usually they dart erratically in swooping dives high in the air. Cradling their legs in the form of a loosely woven basket they ensnare gnats, mosquitoes, and their like in mid-air. Recalling fossils I had seen of ancestors of these ancient insects with thirty-inch wingspreads, I wondered what prey fell into the clutches of those giants millions of years ago.

Some of the red dragonflies were courting. Others appeared to be laying eggs, but there was no water around, and everybody knows that dragonfly larvae live in ponds and puddles. Only from the spring thaw until late June was our lawn one vast puddle. The rest of the year, as now, it was a proper lawn, exceptionally lush because my uncle who had been an avid golfer had sown golf-green seed here to provide a putting green.

It was over this lush grass that the dragonflies were going through the motions of egg laying. Some females bobbed along at grass-top level rhythmically dipping their abdomens with the quickness of a person testing a pan of hot water with his finger. Others flew in tandem—males with their abdominal claspers around the females' necks—in an embrace that seemed more acrobatic than caressing. Thus linked, the pairs also bobbed with the same dipping of abdomens.

But were the females laying eggs? So intent were the couples that I was able to creep close enough to discern that with each dip an elongate cream-colored egg was dropped among the grass stems. This behavior seemed erratic to say the least. Insects often make mistakes in selecting places to lay their eggs. Dragonflies are known to mistake the shiny surfaces of smooth asphalt roads for water, but here there could be no mistaking the lawn for water. Eggs dropped there would lie unattended for months, vulnerable to desiccation, to spiders, ants, and other marauding carnivores. Not until spring would there be water for the young aquatic larvae. Perhaps these eggs were destined for destruction. Yet the presence of the red-jeweled adults proved that some eggs somewhere survived.

Many other creatures caught my attention: a silver-spotted fritillary also laying eggs in the lawn near small violet plants that grew here and there in patches; a black and white warbler exploring the trunk of one of the old white pines; a Maryland Yellow-throat in the brambles, two ospreys crying together as they circled so high as to be mere specks. The whole world of nature simmered with activity. None of these, however, was an inhabitant of our house, but, as I soon discovered, there was another world in our cedar posts, a thriving busy community.

This back porch was long and narrow, more an indentation in the house than a porch, a loggia in fact. Its most prominent feature was a large cedar post with its cedar cantilevers. Two black iron pots of obelia and fuscia hung from the beam. Not even the blossoms of the obelia were too insignificant to occasion visits from the hummingbirds and bumblebees. The late sun picked up minute details of the post. It revealed a halo of fungal fruiting bodies, British soldiers (*Cladonia cristatella*) and other lichens. In particular it silhouetted a busy traffic of small wasps.

The wasps claimed my attention as had those at the chimney, this time by persistently exploring my head. At first I had taken them to be blackflies, but they were examining, not biting. When I sat down my head had become an obstruction to be investigated. Eventually, however, they became habituated and returned to their business at the post. But what was that business?

Scattered throughout the post were numerous small round holes. The symmetry of each was astonishing. While growth and development may provide perfect organic symmetry—as in the arrangement of petals in flowers, the circular pill-box shells of diatoms, or the spheres of many forms of plankton—the construction from inert material of a perfect circle or square is not an easy accomplishment. At the molecular level it is done in crystals. It is quite a different story in the biological world. Try to fashion a perfectly round pot without a potter's wheel or a perfect square or cube without a framing square or other tool. The makers of the holes in the cedar possessed no tools. Their heads, swinging in arcs on the pivots of their necks, were the only measuring tools. But who were the builders? Were the wasps boring the holes or were they appropriating some other creature's tunnels? When I pried carefully with a jackknife I discovered that nearly all the holes had been constructed by powder-post beetles boring from within.

Although my attention had been caught first by minute wasps, I soon discovered that there were two other species, one larger, the other intermediate in size. It struck me at that moment that everywhere nature has its multiple worlds scaled up and scaled down, from the atom with its nuclear sun and electron planets to our own solar system with its planets. In every size category, from the universe beyond us to the universe beneath us, there are patterns of similarity. This principle was exemplified here even in such a banal thing as a cedar post. The three sizes of wasps appropriated corresponding holes of three sizes, sought prey of three different sizes, and were preyed upon by predators of three sizes.

The largest wasps, which resembled smaller models of the mason wasp of the rocks, exhibited much the same habits. Unlike them they were rather particular about the size of the hole selected. They chose holes that very nearly matched the diameter of their heads. Furthermore, they were not satisfied with a bare hole. They first removed debris and pellets of solidified resin before they laid an egg. This done, they began the hunt for aphids.

When the provisioning was completed, the wasp shifted to another mode of behavior. She no longer hunted aphids; whatever had been attractive about them and the leafy pastures in which they grazed now lost its appeal. Clay pits became attractive, but once the hole in the post had been sealed that attraction waned. The entire sequence of behavior beginning with exploration commenced again.

Only one kind of interference seemed capable of upsetting the program. If a wasp was captured in a small glass vial, an easy thing to do because the vial could be clapped over her as she was about to enter a hole, she almost immediately dropped whatever she was transporting, mud ball or aphid, and directed all her energies toward escape. When she was allowed to escape, it was a long time, if ever, before she resumed work at that hole.

The medium-sized wasps appropriated medium-sized holes and the smallest wasps, the smallest holes. The latter were the most interesting, as much for their minuteness as for their activity. Here were no giants, no cicada killers, but, nevertheless, from a thrips's point of view, they were ferocious predators. As with the blood-sucking no-see-ums, the packing of such intricate behavior into such small bodies must excite wonder. If God numbers the hairs on our heads and no sparrow falls unnoticed, how many insects must he be overseeing!

These Lilliputians buzzed around my head for minutes at a time before settling at the post. Even then there appeared to be a great deal of lost motion. A tremendous amount of time was spent exploring. A wasp would come to a particular area of the post, fly back and forth as though scanning the area, and finally land. No sooner had she landed than she would commence running zigzig, all the while palpating with her antennae. Her attention was not even distracted by a small jumping spider for whom the post was a happy hunting ground. Some holes were ignored as though they did not exist, others elicited a brief pause, but others brought the wasp to a complete halt. Closer examination revealed that these select holes were adorned with small circular chimneys constructed of pellets of sawdust. This discovery indicated incidentally that even though powder-post beetle tunnels might be appropriate the wasps were skilled enough carpenters to rebore and renovate to their tastes. And some subtle scent emanated from the extruded material. When I transferred sawdust from an attractive hole to an unattractive one, the wasps became interested in that one.

Attractive holes were explored thoroughly. The wasp would crawl boldly into the darkness, emerge immediately or remain longer than my patience lasted. During one of these long intermissions a neighbor came over with the mail (we had to pick up our mail in the village each day because there was no rural delivery). She asked what I was doing. It must have seemed extraordinarily odd to observe me staring fixedly at one of our porch posts.

"I'm watching these little wasps that are nesting in the post," I explained.

She put on her glasses for a closer look.

"Won't spray get rid of them?" she asked.

"Why yes, of course, but they're really not doing much damage."

"I'd get rid of the pesky things."

I tried to explain that despite their small size they were doing some good in keeping down aphids and thrips.

"Well," she replied, "I don't see how you can stand them. Shall I bring the mail in the house?"

"Thanks, don't bother, I'll take it."

She shook her head in puzzlement, and went around to the front of the house to talk to Lois. I resumed my wasp watching.

There was little doubt that freshly chiseled sawdust, or some secretion such as saliva, provided an olfactory cue, but did this cue enable a wasp to locate a hole topographically or only to identify it as an active site? Two simple experiments settled that issue. First, I pinned a fragment of white paper about two centimeters away from a particular hole and left it there for a day. Its presence did not visibly disturb the wasps. The following day I moved the target to another location about twenty centimeters distant. Now the wasp returning from a flight landed close to the paper, as before; however, not finding her hole, she did a lot of searching on foot. As might have been expected, she knew and remembered her neighborhood and its landmarks from visual cues.

Another wasp that obviously depended very much on vision was one that was using a nail hole in the door trim as a nest. She seemed to have trouble distinguishing nail holes from nail heads because frequently she flew directly to the latter and then had to reorient herself.

Knowing that Hymenoptera possess an acute sense of smell, I performed a second experiment. It was a modification of the first, designed to learn whether or not precise pinpointing of the hole was aided by scent. To this end two targets were pinned on the post, one two centimeters from the hole, the other twenty centimeters away. The near target was scented with citral, a compound producing a lemonlike odor. It did not interfere with normal activities. The next day I switched pins so that the scented target was the far one. The wasp was not fooled. These experiments were not conclusive (perhaps wasps hated citral), but they were suggestive, and perhaps another time in another place some inquisitive person will take up the challenges.

There are many unsolved mysteries in the homing behavior of the wasps. I leave you with the rocking-chair mystery. On the front porch there is a rocking chair, also constructed from unbarked cedar. The legs, rungs, and back posts have their population of little wasps as well. The mystery is: How do the wasps find their way back to the chair since it is frequently moved from one location to another?

5 ○ SILKEN FLIGHT

IN the village, indeed, everywhere back from the coast, the day had been oppressively hot. Even the birds had taken an afternoon-long siesta. The only singer, except for a few field crickets luxuriating in the shaded crevices of an ancient stone wall, had been the "hot bug," the cicada. All day long towering cumulus had marched eastward only to vanish offshore as they passed beyond the outer islands. The wind, fetching out of the northwest, had blown warm and capricious, scattering catspaws over the surface of the bay so that sailing that day tested a skipper's skills.

With sunset the wind died. So calm had the atmosphere become that not even the poplar trembled. The last rays of the sun, itself hidden by the coniferous forest behind the house, picked out the sails of a becalmed sloop and painted them startlingly white against the gathering twilight. Between the boat and the shore the head of a seal appeared hardly distinguishable from the lobster buoys. In the ever-deepening twilight some moths had already ventured from their resting places. A persistent dragonfly still hawked, making one wonder how it could possibly pick out its prey in the dim light. It dived for one moth and caught it. Within a matter of minutes the light became too feeble even for a dragonfly, but the moths gained no respite by its departure because its patrol was taken over by a bat in a nice Box-and-Cox sharing of the hunting preserve.

In less than an hour the temperature dropped a full thirty degrees. Not the slightest breeze disturbed the quiet, but a steady flow of cold air, a silent unseen river, moved seaward. To the west the sky retained a mere suspicion of lightness. Jupiter, with Saturn above, completing its transit of the sky, was setting rapidly over a horizon of trees, trees that were neither green nor black, only very dark. Directly opposite to the east, a full moon, loomingly large and orange, rose over Mount Cadillac. It cast a broad path of light across a flat bay. The mirrored surface moved only by the rising tide.

Some nights are bottomless black, the absolute in blackness. Some nights are soundless, so silent that one can hear the silence. This was such a night. One could strain and strain and hear nothing but the silence. You may argue, as a high-school English teacher of mine once did in criticizing a youthful composition, that to hear silence is a contradiction in terms. Both poetically and scientifically she had failed to understand that real silence is more than the absence of sound. One becomes conscious of the random movement of air molecules, Brownian movement, banging on the ear drum, punctuating the silence with a sensation that seems to come from within.

Such deep silence is a rare thing, and later as I lay in bed reflecting on the daytime chores completed and those put off till another day, the night came to life. From the damp grass that edged the flower garden the Carolina ground crickets, lovers of cool, moist vegetation, began arias that would last the night, each a continuous, weak, nonmusical, metallic, stuttering vibration. Out on the bay two loons began a conversation. Then a slight shift in the flow of cool air caused a vagrant breeze to well out of the west. This finally set the poplars to trembling so that for a moment I was deceived into thinking that rain had begun to patter on the roof.

The interior of the bungalow seemed to be stuffed with a palpable darkness accentuated periodically by the flashes of a firefly that had wandered in from the field. That firefly lent a friendly atmosphere to the blackness and gave it a dimension. It would also have comforted anyone who feared the dark. How could there be any evil where such a cheery lantern glowed? Still, between interruptions in the gentle night noises outside, the silence within was profound. Had it not been so, I would have missed the strangest noise of all.

In this silence, at some point in my recollections of the day, there came upon me an uneasy feeling that we were not alone. The rest of the family was sleeping soundly. Nothing stirred, but then again something was stirring, something that failed to alert the dog, Gulliver.

As I strained to hear, I became conscious for a fleeting moment of the faintest of sounds resembling a fluttering heartbeat. It was not that of my own heart because it came to me from outside myself. For a second or two it sounded just over my head, and very close. As quickly as it came it was gone. Within another few seconds it came again. Was there also a whispering of air? I could not be certain.

Superstition aside, there was only one creature that could account for these perceptions. In this manner did the little brown bat announce its residence. I lay awake awhile longer listening to the silken flight of this truly remarkable, much maligned, much persecuted master of space and darkness. Among millions of manifestations of life, from viruses to ourselves, only the insects, the birds, and the bats have completely conquered the air, and when night falls, the bats and moths have the skies to themselves. In this darkness they play with consummate skill a deadly game in which each stakes his existence. If the moth loses, it forfeits its life to the appetite of the bat; if the bat loses too often, it risks starvation.

In the darkness, hearing and smell are the only senses that count. The bat listening for echoes of its hunting cry bouncing back from the body of the moth zeroes in for the kill, while the moth, hearing and locating the direction of the cry, takes violent evasive action. And although my ears were not tuned to the sounds of the hunt, I knew full well that the deadly game was being played out that night in the bungalow as well as out of doors because the species of moths that bats pursued were no strangers to the bungalow.

To realize that a whole world of life and death is being enacted before a person's eyes and he cannot see it, and around his ears, and he cannot hear it is to diminish whatever feelings of superiority and arrogance one may have. It is to feel humble in the knowledge that there are other worlds and other perceptions. It is to appreciate that we are surrounded with an infinity of wonder and beauty.

The little brown bat knew the interior of our house more intimately than any of us. It avoided every obstacle, each rafter, each piece of furniture, the hanging ox yoke that held two lanterns to light the living room, the iron rod across which the wood mice ran, as well as us lying in bed. As a sailor I appreciated the bat's navigational skill. I knew firsthand the anxiety and tension of navigation by dead reckoning through the pea-soup fogs of Maine waters when the bow of the boat was invisible from the stern. I was humbled by the superiority of the bat, the more so since no sailing vessel traveled at even a fraction of the bat's speed and no skipper routinely had to make such split-second decisions for changing course.

Secure in the knowledge that a navigator of supreme skill was cruising the darkness around us, I fell asleep without the slightest fear that this voyager would be so inept as to become entangled in the hair of any of us, despite what the old superstitions averred.

The following morning the four of us had an opportunity to meet our nocturnal housemate at close quarters. Lois discovered him jammed between the front door and the screen door. Actually he was neither jammed nor trapped. Little brown bats are so very small that they are capable of crawling into the narrowest of crevices. They manage to squirm easily through cracks as narrow as one half an inch. The confined space between the two doors was more than ample.

How he came to be in such a conspicuous place is easily explained. Since he navigated in darkness and presumably had sought shelter before dawn, he had had no way of knowing that the comfortable space he had chosen was bounded on one side by a transparent screen and on the other by a pane of glass.

"Get rid of him," said Lois.

"Let's keep him as a pet," said Jehan.

"Turn him loose," said Paul.

"Bats don't survive well as pets," I reminded Jehan. "When they get over their fright, they learn to take insects offered by hand and to drink water, but for some reason, possibly lack of exercise or boredom, they never live very long."

The bat's appearance was not such as to evoke empathy. Like the wood mice he had large ears, but they were more pointed, and that in itself struck an alien chord. His very small eyes did little to enhance his appearance. Furthermore, in fright he bared a set of needle-sharp teeth. When handled he was extremely aggressive. While in flight he was a thing of beauty, but when earthbound he appeared as a crippled and deformed little monster. Closer examination revealed that "he" was a female.

This revelation suggested that our bungalow probably harbored a maternal colony. Little brown bats mate in late summer. Sperm are stored by the female all during the long winter hibernation. At the end of winter females disperse to their summer homes where they gather into maternal colonies, give birth, and raise their young. When I explained this to the family, there was a consensus that we release the captive. Neither this female nor any other intruded further on our privacy, but now that we were aware of the presence of bats in the area we saw them frequently and wondered how we could have missed them before.

Even though a colony inhabited the bungalow, we never saw it as such; that is, there were no rows of bats hanging from the rafters or packed masses of bats in chimney corners. They carried on their housekeeping discreetly out of sight in external crevices in the roof structure.

We did, however, often surprise one or more at their roosting sites. The most conspicuous sites were the eaves and rafters of the open front and back porches. Between periods of hunting the bats would return to these places to digest their catches, the evidence thereof having to be swept up each morning.

Occasionally late at night we would see one of the hunters hanging from a rafter more or less directly above the porch light. In the tropics I have observed lizards that had learned to hunt around lights because of the bounty of insects attracted to them, and at first I thought that the bat might also be taking advantage of the situation. This did not seem to be the case. The moths and other insects that flocked around the light were safest from the bat the closer they were to the light. Bats, being aerial hunters that screen wide sweeps of unobstructed space, foraged only beyond the outer circle of light.

On moonlit evenings we felt a proprietary interest in "our" bats as we watched from the porch their acrobatic foraging, their swoops, dives, stalls, and turns pivoting on a wing tip. If we threw a small pebble in the path of flight, the bat would abruptly swerve toward the pebble and dive at full power as the stone fell, only to veer away at the last moment. The response to a tossed mealworm or pill bug was clearly different. Obviously, the light was too dim for us to observe exactly what occurred; however, the pattern of flight showed no last-minute veering away from target. Another bit of circumstantial evidence suggested that the bats could indeed differentiate between pebbles and genuine prey. If we repeatedly fired pebbles at the bats, they ignored them after four or five disappointments and eventually moved to another hunting area. When we lofted pill bugs or insects of sufficient weight to be thrown, the bats remained in the area until we tired of the game.

It was certainly snobbery on our part to assume that the local night sky was populated only by bats from the bungalow. Not every house is appreciated by bats, but just what it is about a house that qualifies it as acceptable is something known only to them. By sheer coincidence we learned that our neighbor's house also qualified. The information came in a telephone call for help.

"Come over right away, please. Please! There's a bat in the bathroom." Her tone sounded a shade hysterical so I hung up and hastened over by way of a short cut through the woods.

"Oh, I'm so glad you came," she said.

"Where is the bat?"

"In the bathroom," she repeated, pointing but making no move to accompany me.

Once in the bathroom I looked about but saw no bat.

"It must have escaped," I called.

"No, no! It's dead."

"But where is it?"

"In the toilet."

I lifted the lid, and there indeed was the bat, drowned. I removed it, a little brown bat, and told my neighbor that I would dispose of the corpse. I did not have the heart to suggest that there was probably a colony in her house as in ours.

The circumstance of the drowning was not so mysterious. Bats, like all animals, must drink. Life under the roof or in an attic undoubtedly gets uncomfortably hot during sunny summer days. This heat stimulates a great thirst. It is not improbable that this bat detected the water in the toilet bowl and in attempting to slake its thirst lost footing on the slippery porcelain.

The bats, like ourselves, were strictly summer visitors, and shortly after Labor Day they joined the general exodus. On occasion when we returned for brief autumnal weekends during Indian Summer, there was never a sign of them. They may have migrated to a special cave in the Green Mountains of Vermont where many New England bats congregate. Some go to the cave from as far away as Cape Cod and parts of New York State. Bats banded in that cave in the wintertime have been recovered in their summer homes and then again in the cave during the following winter. It would be no unusual feat for our bats to make the trip to Vermont, but since they had not been banded, we had no way of knowing whether that cave or some other, nearer at hand, served as their winter residence.

How they find their way between winter and summer homes is even more of a mystery than is the navigation of birds. The distances involved may be great. One record was posted by a little brown bat that returned home from a distance of 270 miles. It is likely that bats possess some sort of a compass for distance navigation and also recognize certain land marks, especially when near their destination, by echolocation and by vision.

These skills which so far exceed our own in these domains, together with our inability thus far to fathom the mystery of their accomplishments, could not but bias favorably our attitude toward the small mammals. Naturally a houseful of bats is quite a different matter than one or two, but then so also is a crowd of people compared to a small intimate gathering. In the final analysis, if that single bat had not spent one night foraging inside the bungalow, we would never have become aware of its occupancy. Except for slight indiscretions on the outside porches the bat was the most private and discreet of all our guests.

6 ° A COMMUNITY OF ROBINS

T H E rooster, fabled Chanticleer, is a fraud and a laggard. The first bird to awake in the morning is the robin. In truth, he begins to call before the sky pales in the east. It is probably he who wakes the rooster. Years ago as a youngster on Cape Cop I was being allowed to go with my parents on an expedition for deep-sea clams. These could be obtained on certain offshore bars on which the water was shallow enough only at the lowest moontides. On this occasion we had to start out at four in the morning, and, like a child, I kept waking up every hour on the hour all night in anticipation. I know the robin is first. More recently in Maine I was able to verify this observation on several occasions when one thing or another got me up before dawn.

It is fitting that the robin should occupy this position of primacy. He is probably the bird traditionally and emotionally closest to human beings. Part of this feeling of companionship is undoubtedly historical. When early European settlers observed the robin, they were reminded of their robin back home, but the English cock-robin is a different and unrelated bird; our robin is really a thrush. Nonetheless this red-breasted thrush became a beloved surrogate.

This historical sense of companionship was reinforced by the rapidity with which the American robin adopted human beings, their houses for nesting, their lawns for gardens for feeding. Our bungalow was no exception.

To appreciate the haven that the bungalow offered for nesting birds you must be able to visualize the construction of the front and back porches. The cedar posts that were inhabited by the small wasps supported a ten-by-ten beam which in turn supported the rafters that slanted down from the peak of the roof. These rafters projected nearly three feet beyond the beam to provide unusually wide eaves. The outside surface of the horizontal beam was shingled in such a way as to provide the porch side with a series of cubbyholes between each pair of rafters. For as long as I can remember robins had adopted these boxlike compartments for their own.

Each year a variable number of robins nested on the two porches. One summer three built nests on the back porch. Of these, two built on the beam and one on the telephone and power lines where they were fastened to the house. Every spring I have the greatest difficulty preventing the servicemen from knocking down that particular nest.

While I find a cloying sentimentality over animals distasteful— although I can forgive it in children—the opposite extreme of complete indifference toward inoffensive and vulnerable animals is incomprehensible. The nest in question did no harm. It did no violence to the aesthetics of the bungalow; the nestlings, unlike those of swallows, were not messy.

The actual number of nests on the beam was six. At first I thought that they represented the work of six robins. Closer attention revealed, however, that each of two robins was building multiple nests, some only partially completed. A cubbyhole on a beam is a most attractive nesting site. When by the repetition of motifs architecture produces multiples of attractive nesting sites, in this case a series of cubbyholes, the robins are overcome by the alluring plethora and are carried away by a frenzy of building. Under natural conditions some birds, as for example, wrens, build dummy nests, and one can appreciate the survival value of this practice. It is not, however, in the nature of robins to do this. Only when confronted with

what may be a series of difficult decisions do the birds vacillate between this site and that.

Nests on the beam were out of sight. They were constructed in a peculiar manner adapted to the situation. A casual observer is inclined to view a robin's nest as a rather untidy, bulky, unitary structure; however, on analysis it is seen to consist of a symmetrical mud cup set in a loosely constructed foundation of grasses, weed stalks, and slender sticks. The cup itself is neatly lined with fine grass. The foundation is artfully formed to fit the contours of a tree crotch or a horizontal limb that it may saddle. Adaptation to life on a flat beam resulted in the nest foundation being built in the form of a thin flat mat. The cup was set on this mat rather than in it, as was the usual relation between cup and foundation. Despite this aberration in style the nests were carefully crafted.

Unlike house construction there is no shoddy workmanship for the simple reason that poorly constructed nests lead, through destruction by weather or accident, to the demise of the eggs or the young and the consequent elimination of "shoddy-construction" genes. Birds that are poor builders, inefficient foragers, bad parents, or just plain careless of their own lives leave no offspring to perpetuate their deficiencies.

Year after year there were robins' nests on the porch beams. Since the birds were not banded and had no distinctive individual characteristics as far as our unbirdlike eyes could discern, there was no way of knowing whether the occupants were the same individuals each succeeding summer. We had evidence from other sources that individual birds did tend to return to the same sites in successive years. One summer, for example, one of the boys, while giving the lawn its first mowing of the season, flushed a song sparrow from her nest in the tall grass. Within a foot of the nest we discovered another, older nest. The following year a song sparrow built still another nest near the first two.

Since one part of the lawn, really more of a hayfield than a lawn, looked much like another part to our eyes, it seemed unlikely that three different song sparrows had chosen precisely the same spot in the broad expanse of waving grass. We discovered a comparable situation with chipping sparrows. There were three successively aged nests in an isolated barberry bush at the edge of the woods. There was a curious feature of these nests that at first escaped my notice. It had to do with the lining. Characteristically chipping sparrows line their nests with horsehair. At least this was so before the automobile replaced the horse, and horses were common throughout the countryside. I was a bit surprised, therefore, to discover horsehair linings in these nests, because the nearest horses were in the next village. More careful scrutiny revealed, however, that the "horsehair" was nothing of the sort. It was very fine strands of some kind of electrical insulation that in texture and appearance totally resembled horsehair. This substitution suggested that the sparrow's visual perception was very similar to our own. That which resembled horsehair to us also resembled it to chipping sparrows. Is this to be wondered at? Perhaps not; on the other hand, we know that in many matters animals see the world differently than we do, and we can never take for granted that all of us see the world the same way.

In the wild, most song birds are rather short-lived. The greatest recorded age of robins in their natural environment is ten years, but the majority probably survive for only three to five. Clearly, therefore, successive generations of robins were populating our porches. It would have been interesting to know whether they were lineal descendants of the first robins to have discovered the bungalow or whether some special attraction invited them.

Unlike chimney swifts, swallows, and phoebes, which nested on cliffs and ledges before New England was colonized by Europeans, robins have always nested in trees. Special features of houses offer some of the characteristics of cliffs so the transition from one to another does not involve a drastic change in style of living for swifts and swallows. With robins the rationale is not so obvious. Yet within a period of less than three hundred years robins have come to accept houses as desirable nesting sites.

A house does offer unique protection against predators. Even an unoccupied house or barn must evoke a sense of claustrophobia in those predators that hunt in boundless spaces. Add to this the presence of human beings in an occupied house and the protection must be nearly complete.

In our area we had seen three predators of nesting robins at work close by the bungalow. The first encounter took place in a venerable white pine growing in the open between the bungalow and the well. This pine was a particular favorite of robins. It also offered nesting sites each year to a chipping sparrow, a pewee, a magnolia warbler, a red-eyed vireo, and a cedar waxwing. One year a goldfinch chose it and another year, a black-throated green warbler. From these tenancies you can imagine that this tree is a giant, as indeed it is. Old sepia photographs taken at the time that the bungalow was built show this pine to have been enormous even then. Thus I know from hard evidence that the tree is more than seventy years old.

Indirect evidence in the form of a rather unpleasant experience enabled us to estimate that the pine is close to a century and a half in age. The evidence was provided rather suddenly when a pine of the same girth just behind the house was struck by lightning. Even without that particular bolt the storm was frightening. Although the time was four in the afternoon, the sky darkened so deeply that we had turned on the lights. Thunder pounded ever closer as the storm rushed upon us from the west. Jehan was counting intervals from lightning flash to thunder clap.

"One, two, three, four, five—" Crash!

"It's a mile away," he remarked.

"Where are the candles," shouted Paul above the din of two more crashes that caused the telephone bell to tinkle. We knew from experience that there were two out of three chances of a power failure.

"Light one of the hurricane lamps," I suggested. "Where is your mother?"

"In the kitchen."

I hurried there and shouted—I had to make myself heard above the crashing and pounding—"Lois, don't touch any of the faucets."

"One," began Jehan. He never finished. The most tremendous explosion of light and sound enveloped the bungalow. It appeared, as we all agreed later, as though a huge fireball filled the living room. Surely the house had been struck. An acrid odor of burning electrical insulation came from the boys' room and the kitchen—a lamp cord in the former and the washer and dryer in the kitchen. But the bungalow itself was intact and no fire started. The bolt had struck instead a huge white pine in the back yard. It had split it cleanly from top to bottom and had scarred two adjacent white spruces with long ragged brands. Through the greatest of good fortune neither the pine, the spruces, nor the bungalow caught fire. The next bolt flashed out to sea as the storm passed southeasterly. As predicted, the power failed; however, the telephone still worked. A cousin who lived on a hill in the direction of the village had seen the lightning strike.

"Are you all right?" he asked. "Did it hit the bungalow?"

I described what had occurred.

"Well, we know it was close. We saw a spectacular orange flash and were concerned about you."

When the storm passed, as rapidly as it had arrived, we trooped out to examine the tree more closely. It was a wonder that such a resinous tree had not instantly burst into flame. The heat had been so intense that the wood was as seasoned as if it had been kiln-dried. One has no realistic conception of the immense power of lightning until he has witnessed its effect firsthand. Here a tree that was four feet in diameter had split like a match stick.

A few days later the boys and I felled the stricken giant. Then it was that I counted the rings. The tree was one hundred and forty-eight years old, give or take five years. Thus we arrived at an estimate of the ages of a half-dozen other white pines growing around the bungalow.

Viewing the felled one in the forest I began to see the bungalow from another perspective. As the house had its characteristic populations and communities so did the great white pine. Some features the two had in common; others were unique to one or the other. The predominant feature of a pine was its living nature. Its massive trunk offered living wood whereas the stuff of the bungalow was dead. Carpenter ants and powder-post beetles could not live in living wood; insects that feed on it inhabited the pine.

Its needles offered living tissue in a form attractive to caterpillars, sawflies, aphids, and myriads of insects that would die of starvation in the bungalow. The white pine was parasitized; the bungalow could never be parasitized. As an animate unity the pine attracted its parasites which in turn attracted predators upon which still other predators feasted. In this way the tree provided food where the bungalow was barren, and hazard where the bungalow was secure.

As a refuge the pine offered no fewer and probably no more diverse features than the bungalow; it simply presented different ones. It mimicked neither cliffs nor ledges, caves nor crevices, damp puddles nor dry crannied rock. It provided needle-shielded forks from which the red-eyed vireo could hang her nest, sheltered terminal branches at low levels where the chipping sparrow could conceal her hair-lined cup, loftier spreading branches in which the magnolia and black-throated green warblers could nest in high safety, platforms along the mightier, lichen-encrusted horizontal limbs where the pewee camouflaged her nest in the open, hanging bunches of Usnea moss within which a parula warbler could weave a nest, and everywhere appropriate niches for waxwings and robins. Specific diversity and form, and, above all, life characterized the white pine.

For robins it did, nevertheless, pose some hazards that the bungalow minimized. In the tree near the well a robin had built a nest fitted comfortably half way along a limb that stretched over the driveway. At the time I did not know whether it contained eggs or nestlings, but some other eyes were more observant than mine.

I had noticed that crows now and again perched in the top of the pine, but I paid little attention to them. Apparently one had kept watch on the robin. One morning as I came around the corner of the bungalow I surprised this marauder in the act of removing an egg. Startled by my appearance it took off with the cracked egg, dripping yolk, gingerly grasped in its beak.

This incident substantiated what I had already known, namely, that among the items of a crow's diet birds' eggs were commonly chosen. Bold as this crow was, however, it would never have come to the porch to vandalize the nests there.

Another predator, admittedly not a usual one, harassed a robin that had chosen to nest in a gnarled apple tree near the brook. Years ago, before the forest had marched upon it, an orchard had flourished there. Not far along the same brook a broad-winged hawk had built a huge but compact nest in a crotch of the main trunk of a very old white birch. This hawk tolerated no intruders in its territory. It did sentinel duty from the spire of a tall red spruce. When we ourselves crossed its sovereign boundary, it circled overhead with piercing, threatening cries. If Gulliver hunted around in the underbrush, it became even more aggressive. When the erring robin attempted to approach its own nest, the hawk pursued it, harassing it so much that this bird finally abandoned its nest.

The third predator whose marauding I had witnessed on several occasions was the red squirrel. Although these scamps feasted for the most part on seeds from spruces and balsam cones, they seldom hesitated to add animal protein to their diet when they discovered a nest of eggs or young. They were less afraid of houses, and have even been our tenants, but a nest on the porch was much safer than one in a tree. Thus, for the robins, the bungalow was ecologically very desirable.

In accepting our hospitality the robins made certain adjustments. The model of the nest was one. The multiple nest building was another. A third was their compromising the dimensions of their territory. Sociable or colonial birds may breed very close together, but even then there are definite territorial limits. Solitary birds stake out territories of some considerable extent from which they warn intruders by singing their proclamation of songs from treetop, housetop, and fence post. The robin in the big white pine tolerated no other robins in that tree. On the back porch the occupied nests were no more than a few yards apart. These robins had forsaken the countryside to become urbanites. The extremes to which urban robins will adjust are unbelievable. Among numerous cases reported is one in which a robin built a nest on a moving ferry and traveled back and forth with it, feeding the young and successfully bringing the brood to maturity.

From the robins' point of view we were the intruders. In one sense they were correct because they had completed building their nests before our arrival in late June. Territory belonged to him who got there first. Of course it had to be defended. The robins' defense took the form of agitated cries and considerable flying about, but there were no actual or feigned attacks. In time, since we left them strictly alone and even avoided slamming the screen door, they became accustomed to our comings and goings.

Generally speaking life passed tranquilly enough for all of us. The robins were tidy housekeepers. Faithfully the parents picked up each little bag of refuse, each little membrane-wrapped sack voided by the young, and carried it off to the woods for disposal. In this manner the nest was kept clean, thereby reducing the incidence of disease. Additionally the location of the nest was not betrayed by the appearance of a privy on the ground immediately below.

In lieu of rent the robins wormed the lawn, by which I mean they did their utmost to keep it clean of grubs, caterpillars, grasshoppers, beetles, and earthworms. I begrudged them the earthworms whose industrious tilling I applauded, but I figured that the earthworm population could survive the annual cropping.

The aesthetics of the presence of these robust birds on the lawn was a constant source of pleasure, especially in the early evening when we relaxed on the screened side porch to savor the peace and beauty of the passing day. Surrounded with forest as we were, except on the seaward side, we saw no signs of human activity unless a neighbor came to call or a tradesman to deliver. Boats on the bay provided a sense of contact, an awareness that we were part of a greater whole. The robins brought life closer to hand. They provided companionship in a way that a scene without movement or a painting does not. They reminded us that the world did not exist for us alone. Going about their normal workaday business they seemed to be telling us that all was well with nature. They also gave us a sense of being a part of nature and having, therefore, a responsibility of stewardship.

7 ∘ ON LONG POINTED WINGS

Y EARS ago before telephone lines were gathered into cables and cables were buried underground, long lines of telephone poles followed every country road. Though the roads might have been devoid of travelers, the mere presence of the many strands of thin wires strung from pole to pole gave the assurance that somewhere along the road, around the next bend, over the next hill, or beyond the forest, was a town, village, or house. And where there was one of these, the wires would, at appointed times of day, be strung with swallows as surely as clothespins on a clothesline.

Telephone wires and swallows seemed designed for each other. The fine, graceful lines of the wires etched against the sky lacked completeness without the swallows. No other birds would do. No other birds possessed the grace and elan to match the wires' sweeping harmony, delicacy, and purity of line. The swallows and telephone lines belonged to each other as notes and staffs.

In yesteryears, early in the morning and late in the afternoon, hundreds of swallows on telephone lines were common sights. On days when fogs kept insects and birds alike out of the skies, the lines were crowded more densely than ever, one above the other and from pole to pole. The homely, neighborly feeling that this sight evokes is missing now that farms have been abandoned, tired barns are collapsing in the weakness of senility and neglect, and the wires are gone.

Swallows, among the most aerial of all land birds, do not actually need the wires, but they need what the wires signify in open rural country; they thrive in a land of fields and farms. For this last reason our bungalow is not ideally suited for swallows. Although the bank at the beach, where the kingfisher once nested, was also host to a pair of rough-winged swallows for one year, that species is the least common of the local swallows. For bank swallows there is too little space to accommodate a colony. Only in the next town at the site of a large gravel pit is there room. The forest presses too closely on the bungalow for other species to be comfortable.

Around the headland in the village of East Bluehill three species have found a congenial environment—fields, pastures, small open ponds, shelter from storms, and most important, clusters of buildings. Barn and cliff swallows find an abundance of nesting sites in the few barns still standing and under the eaves of the houses, the grange hall, the miniature post office, the equally miniature library, and the sheds of the boatyard. Where people erect bird houses, the tree swallows are also abundant.

All together they are an exuberant throng. What a pity that some of the villagers are so unfriendly to the cliff and barn swallows! The cliff swallows in particular have a difficult time because people each year knock down their gourd-shaped nests.

"They are messy," say most.

"They're a damn nuisance," say others.

"They spoil the looks of the house," say neat people.

I've never heard anybody say, "I love their graceful ways. They make me happy as they sweep through the sky. Their twittering is so cheerful." Or even, "They keep down the bugs."

Despite the discouragement, however, the birds keep returning.

It was a special pleasure to us, therefore, when three summers ago swallows discovered the bungalow and, despite its imperfections, decided to rent for the summer. They did not agree in so many words to pay, but we knew they would help to control the insects and would provide cheer.

I suspect that they found us as they explored along the coast. The open front porch facing a boundless sky and sea was the nearest thing to a wide-open barn door. The robins liked the back and side porches because they faced garden, lawn, and woods; the swallows liked the front porch for its open communication to space and far horizons.

The first arrivals were a pair of cliff swallows. We had already settled in for the summer when they began to explore. As soon as I noticed them, all of us avoided that porch as much as possible in the hope that they would build. And they did. On the first day of construction they stuck a few mud pellets on a rafter where it extended from the side of the house. The next day nothing was added. Whatever the reason for delaying, the pause gave the first pellets an opportunity to dry so that they became firmly cemented to the rough unpainted wood. Too much fresh mud applied at one time most certainly would have caused the nest to fall of its own weight before it had an opportunity to dry.

From that day onward the nest was built up pellet by pellet. Gradually it assumed the shape of a cup. Each curving course of pellets was made successively longer until the cup reached its maximum diameter and depth. After that each course became shorter and more sharply curved. This resulted in the sides being raised and drawn together to form a roof. Shorter and still more curved became the succeeding courses as the opening in the roof began to close until the rows of pellets formed a complete circle. This was drawn together and extended partly downward in the form of an incipient funnel. It is at this point that the individuality or experience of the builders expresses itself. Some build rather pronounced funnels facing downward; some settle for plain unadorned doorways; some leave rather large entrances high up on the side of the nests.

It took four days to build the outer structure of this nest. In general, construction time depends very much on weather. Building ceases when the weather is rainy. If the days are very humid but otherwise conducive to building, construction may proceed at a rap-

id rate, but the results may be disastrous because high humidity prevents the mud from setting. It would be interesting to know whether experienced swallows make allowances for different rates of drying. It would also be interesting to determine whether the characteristics of the mud affect construction methods and duration. Sandy soil makes poor natural cement for swallows.

Our area suits mud builders to a tee because of the ubiquity of fine clay. A century and a half ago these clay soils were discovered to be excellent for bricks, and a local brick factory flourished for a few decades. The abundance of blue clays also stimulated the rise of potters in the area. At the present time, however, only the birds and insects continue to use local clays. The brick factory succumbed to progress. The potters now import superior clay for their stoneware.

We did not observe the process of lining the nest with grass and feathers. Nor were we able to observe the eggs and newly hatched young. We knew only that enormous numbers of insects were carried into the nest from day to day. We wondered at the tremendous amount of energy expended in foraging. Even though constant flight did not require a continuous beating of wings, there being much skimming, diving, and looping, quartering the sky for hours at a time is no pastime for sluggards. Like swifts, whip-poor-wills, and nighthawks, these birds screen the air with wide gaping mouths to net the aerial plankton.

Finally one day we looked up and saw two faces looking down at us from the entrance of the nest. Different fledglings poked out of the hole from time to time. The parents no longer entered the nest. There could have been hardly any room. Instead they clung to the edge of the doorway and stuffed flies, mosquitoes, aphids, and other delicacies into the cavernous mouths. Eventually all their efforts were rewarded by five new additions to the resident swallow population, hardly enough to cause a population explosion even when all the fledglings of the area were summed. The toll of migration, predation, and disease holds the balance.

The following summer the cliff swallows did not return. Perhaps the area was not open enough, perhaps, being social birds, they missed the company of others of their kin, perhaps they did not survive the rigors of migration. For whatever reason, we lost them.

Few things are permanent and continuous unalloyed joy can lead to boredom; but we missed them nonetheless. Our disappointment was softened by substitutes. Two pairs of barn swallows had discovered the porch. The summer of their arrival was the same dry summer that had given the chimney-dwelling wasps so much trouble. The small oasis at the foot of the bank could not accommodate swallows. Their demand for mud was considerable and I feared that the lack of it would cause them to desert us.

The obvious solution was to make mud for them. This I proceeded to do to the annoyance of people who drove into our driveway. There was a low spot in the driveway which always collected water during rain storms. In the spring, as a matter of fact, it collected and held so much water that cars sunk hub-deep. Thus, it was a simple matter to make a puddle there.

In no time at all the swallows discovered the mud. Now, just as we had done with the wasps, we could watch the making of mud pellets and the construction of the nest. The fundamentals were the same for wasp and bird. Here were two animals making mud pellets with their mouths. It was amusing to watch these trim birds on their small stubby legs standing at the edge of the pool and pecking and kneading the mud. I wondered if they saw their own reflections against the reflection of sky and clouds and if they recognized themselves.

We who take our exquisitely talented hands for granted, unless we are so unfortunate as to lose the use of them, do not fully appreciate the miracle of hands. Nor do we appreciate the versatility of the mouth and its appurtenances. Mouths, jaws, mandibles, beaks, and maxillae are not just for eating. They are the hands of the handless. What wonders of craftsmanship the skilled mouth can execute!

As I admired the swallows building their nests I suddenly realized the universality of the use of mud. At our bungalow alone it was employed by robins, swallows, phoebes, the wasps on the chimney, and the mud daubers under the eaves. If one considers the world at large, the list can be expanded to include at one extreme the primitive tubes of certain lowly marine worms and, at the other, the magnificent and gigantic castles of tropical termites.

When we speculate about the evolution of construction materials used by human beings, we are reminded of the primitive mud huts of Africa where mud is plastered on a framework of poles, the walled mud city of Kano at the southern terminus of trans-Saharan caravan routes, the imposing ornate mud skyscrapers of ancient Arabia, and the adobe dwellings of southwestern tribes of American Indians. We can wonder whether mankind learned the trick from other animals or whether the idea of mud for building was conceived in a flash of inspiration. But then, before congratulating ourselves on our superior intellect we must consider the birds. Our cliff and barn swallows were building nests of adobe bricks 130 million years ago. The principal difference between them and us was that men sun dried their rectangular bricks before using them while the swallows began construction with undried globular "bricks." Even the birds cannot claim priority, however, because several million years earlier the ancestors of our mud daubers had already invented mud construction.

Our curiosity regarding the number of eggs in the nest could easily have been satisfied by climbing a ladder; however, fearing that this intrusion might frighten the parents, we contented ourselves with waiting until the young were large enough to peek over the edge of the nest. Unfortunately this was also the age when they could poke their little rear ends over the edge to defecate. The parents did indeed pick up the neat fecal bags and fly off to dispose of them, but either they could not keep up with the excretory rate set by the five fledglings or they were lazy. Whatever the reason, the porch below the two nests began to get messy and would have become filthy had we not invented a solution to the problem.

It is usually at this point that human landlords lose all patience and knock down the nests. Some, anticipating the problem in advance, demolish the nests as soon as construction commences. Early dispossession is probably less heinous a crime than breaking the leases after a family has been started. Our solution was a compromise suited to landlord and tenant alike. On wall brackets below each nest we placed a three by three foot platform of plywood to catch the droppings. At the end of the breeding season we removed the catchment for use the following summer.

After the young had hatched we could share the use of the porch with the swallows. In fact, parts of this chapter were written on that porch in those days. On warm summer evenings when mosquitoes and blackflies were not too annoying we would have supper there. We could look out over the bay, and when the tide was full and the water motionless, we could watch our swallows in the company of others skimming the surface, looping, curving, and describing every sort of trajectory that geometry allows.

Some birds evoke our gaping wonder as they soar on motionless wings, rising even higher on the thermals until they become mere specks disappearing into the cumulus. Late in the summer we have seen ospreys and herring gulls circling in this effortless freedom. Other birds excite us with the raw power of their sustained flight, the hawk in pursuit of a songbird, Canada geese in the fall as they drive southward with indomitable strength and stamina.

Then there are the swallows, rivaled only by the swifts, the whippoor-wills, and the nighthawks. They stir in us feelings of excitement and exuberance. Somehow we felt, perhaps in arrogance, that we had in a small way contributed to beauty and happiness by offering our hospitality and sufferance. The robins gave us a sense of earth's companionship; the swallows, an intimation of freedom of spirit.

8 ∘ STILL MORE BIRDS

ONE need that we share with all the other animals living in our house is water. Each of us satisfies this need in his own particular fashion. The powder-post beetles drilling their neat round holes in the rocker on the front porch, grinding exceedingly slow and exquisitely fine like the mills of the gods, manufacture their own water. From the desiccated wood of the rocker they not only extract nourishment but also rearrange atoms of hydrogen and oxygen from this powdery food into water. The small predators, on the other hand—carpenter ants, spiders, and wasps—survive on the juices of their victims. The little brown bat and other large predators drink water. Birds also drink.

In years of drought some animals suffer acutely. One memorable summer we experienced a record drought. We had been without rain for so long that all nature, like a vast sponge, gave up its water to a thirsty, dry atmosphere. Even the brook in the shade of the forest dried up completely. Occasionally in the past it had ceased to flow, but then small disconnected pools remained in bends, behind boulders, and in deep holes under exposed mossy roots. During this unusual summer even the pools disappeared. Sphagnum moss, which could be counted upon in dry times to yield drinking water when squeezed, only crumbled in the hand this year. For most of the birds the situation was critical. Sea birds, which normally drink ocean water because they are able to eliminate excessive salt through their nostrils, were excepted. For the gulls, scoters, cormorants, and others out on the bay, therefore, there was no crisis.

We ourselves had to resort to extraordinary measures to meet our needs. Originally the bungalow derived its water from a small but reliable spring at the foot of a bank on the shore barely above the mean high-tide mark. This spring had two disadvantages: it was nearly one thousand feet away, and this necessitated a long line of pipe to deliver the water; it went brackish when a combination of tide and wind carried salt water over the protecting cistern. Even if taste did not tell us when this happened, the emergence of lively brine shrimps from the kitchen faucet did. Eventually we replaced this source with a dug well behind the house.

During the great drought the well failed, as did also those of our neighbors. Some people arranged for the volunteer fire department to fill their wells, but this often amounted to pouring water into a leaky bucket. We traveled to Blue Hill to obtain water for drinking and cooking from the town spring. That had never been known to fail.

These conditions prompted me to put into effect, for future use, a contingency plan that had lain dormant since the previous drought. This was to equip the bungalow with gutters, downspouts, and rain barrels. Thus we would in the future be provided with an emergency supply of water should the well fail again. Eventually, I might add, we replaced the shallow well with a driven well.

From the local general store I procured several pickle barrels with excellent oak staves, impregnated unfortunately with the odor of vinegar. Repeated rinsing in the ocean leached out some of the residual vinegar, but I counted on aging to complete the task. In any case, the collected rain water was to be used for purposes other than drinking. Just as birds bathed in puddles we could always bathe in the ocean, but unlike the animals in our house we had other uses for water, laundering and dishwashing to mention two.

The first part of the plan called for the construction and installation of gutters. These were built of wood. It was during the process of installing them that I became further acquainted with our feathered tenants.

At first neither robins nor swallows took kindly to the carpentry that ensued on the eaves of the porches. The swallows eventually became somewhat reconciled to the activity, but the robins protested vociferously to the bitter end.

Two other birds were so shy and unobtrusive that I would never have known of their presence had I not accidentally discovered the partially built nest. It was tucked in one corner of the porch roof on the beam comparable to the one that the robins occupied on the back porch, but this nest was totally unfamiliar. It was sparrow sized, made mostly of rootlets, with some grass, and many pieces of shredded bark. I was pleased to note that the bark had not been vandalized from our cedar posts. The lining, which was unfinished, consisted of very fine grass, finer dark brown rootlets, and some hair, probably Gulliver's.

For the longest time I never saw the builders, and I feared that they had been frightened away permanently. If that were the case, thought I, there would be no harm in examining the nest after my carpentry had been completed. With this in mind I gave the inhabitants two days of peace and then climbed up to examine this unfamiliar structure. Perhaps there were some distinguishing characteristics that I had missed, some signature of the builders. And there was. I was astonished, and pleased, to discover an egg. It was very pale blue-green, almost white, and thickly spattered with reddish and purple spots. Here was a clue.

Three days later I obtained my first look at one of the birds. It was a slate-colored junco, the snowbird of winter. But this could not be! Juncos nested on the ground. In fact just the year before I had found a junco's nest in a typical location at the end of our driveway. It was built into the rank vegetation of an overhanging bank of the ditch beside the road. Water rushed by the nest after each rainstorm, and automobiles roared above the nest several times a day. Aside from these distractons the location was just what juncos preferred. I had never known one to build in houses, although I was to learn later that on rare occasions this occurred.

We now had another house guest to observe. The most comfortable arrangement for all parties was for me to remain in the living room, pull a chair up to the door, and watch through field glasses. Despite all our precautions the birds remained shy and nervous during their entire stay, either because they had elected to move into an alien neighborhood or because it was their nature to be extremely shy.

When one lives with birds on a day-to-day basis, he cannot help being impressed with differences in temperament. Even without quite understanding what temperament really means when applied to animals, it is clear that subtleties in behavior, in mannerisms, are characteristic of each species. The juncos were shy, secretive, wary, just what might be expected of small, vulnerable, ground-nesting birds for whom a premium is placed on concealment during the nesting period. The swallows, on the other hand, quickly adapting to our presence, made no attempt to conceal their nests or their activities. They apparently felt secure in the inaccessibility of their nests, their aerial supremacy, and their personal agility. But the robins, whose nests were equally conspicuous and out of reach, seemed more nervous, agitated, and even aggressive, at least in bluff.

On the kitchen side of the house I discovered during my carpentry still another bird with an entirely different temperament. This was a phoebe. There was no mistaking the nest perched on a ledge over one of the kitchen windows where the overhang of the eaves protected it from the weather. It was neat and compact, and its siding, to use a house builder's term, consisted of green moss. Five white eggs had already been laid.

The birds obviously had begun to build before our arrival. At first they were shy and easily frightened. Within three to four days they accepted us as part of the natural scheme of things. Now not even the thoughtless slamming of the screen door alarmed them as we made frequent trips to and from the woodshed. They would fly no farther away than the clothesline.

From this favorite perch they watched us and from it they did their hunting. In this respect, the keenness of vision of these small flycatchers never ceased to amaze me. The ability of the soaring osprey to spot submerged mackerel from on high or the skill of the busy wood warblers whose bright eyes pick out from leafy backgrounds the smallest and most inconspicuous of caterpillars has always seemed impressive. In my estimation, however, the performance of the phoebes surpassed all. Unlike the swallows and nighthawks that swept the open skies, the phoebe focused on minute individual insects. From its swaying perch on the line it watched the air and sky above the garden. Periodically it would dart into space, pick some invisible (to me) insect from the air, and return to its station where it perched nonchalantly as though the whole procedure were no trick at all. Its tail would sway up and down giving the impression that a wind was blowing a tail not too securely fastened.

Sometimes the intended prey, behaving as though aware of being pursued, would twist and dive. Then an aerial dogfight would ensue with the phoebe executing a series of wild acrobatics, sometimes perilously close to the ground. Victorious or not, the bird would return to the line where it rested with the utmost reserve and composure—a trim, conservative, white and olive bird.

It is ironical and sad that this neat, friendly bird, so expert at hunting flying insects, is incapable of protecting its young from parasitic insects and mites. We witnessed a developing tragedy. The five eggs hatched in due course, the parents dutifully fed and cared for the young, but gradually a population of parasites built up in the nest. The young birds became heavily infested, attacked, irritated, fed upon, and weakened. For some inexplicable reason the neatly groomed parents continued their amazing performance of snatching minute insects out of thin air but were unable to delouse the nest and their offspring. We know that birds are good learners and are perceptually alert. On the other hand, much of their behavior is irrevocably built in. Call it innate, instinctive, genetic, or what you will, occasions arise where it can doom them.

The young phoebes never fledged. One day we missed seeing the parents on the clothesline. Upon inspection of the nest I found five desiccated corpses.

Only one other bird considered our bungalow a fit neighborhood. We had arrived unusually early one summer, exactly at the beginning of a vicious blackfly season. None of us was aware of the newcomer's nest-building activity until shortly after sunrise on the first morning when we were awakened by an ear-splitting racket. It sounded as though some maniac with a jackhammer was trying to demolish the kitchen. Half asleep I stumbled into the kitchen to investigate. For a few moments the din subsided but soon resumed with undiminished vigor. Obviously a woodpecker was hard at work on the outside of the building.

Instead of going out through the kitchen door I padded out to the side porch and on to the dew-soaked grass where I surprised an early robin after the proverbial worm. Continuing stealthily around the house, I peeked cautiously around the corner and caught sight of a flicker hammering determinedly at the door trim. At about the same moment it caught sight of me and fled into the woods.

Carefully centered in the trim at about a six-foot level was a freshly chiseled, round, flicker-sized hole. Since I was now fully awake, and the morning bid fair to be a beautiful one, it made little sense to return to bed. It did make sense to get dressed before examining the hole.

The flicker had chiseled completely through the trim and far enough through the underlying siding to have appreciated, if flickers appreciate these things, that this part of the house was hollow. What the bird could not have appreciated at this point was that, unlike hollow trees, the hollowness of the house was deep and extensive. The bungalow had no interior finish. If the flicker had completed its hole, and entered, it would have found itself in the kitchen.

Fearing that this bold woodpecker might be a creature of great persistence I wanted to discourage it from completing its work as well as from making new attempts elsewhere. With this in mind I tacked a piece of tin over the hole. The next morning the din was worse than ever. Far from giving up, the bird was beating a raucous tattoo on the tin.

During courting time flickers seem to enjoy beating out messages on resonant dead trees and tin roofs. "Oh, no," I thought. "What if it gets carried away by the sound, forgets nest building, and begins courting all over again?" Fortunately for our peace of mind, this one was in the nest-building phase. It tried for one more morning to chisel the tin, but then acknowledged defeat. Later that week I wondered as I replaced the trim why the bungalow was so attractive to flickers. It certainly bore no resemblance to a tree trunk.

For that matter, what made the bungalow so attractive to such different kinds of birds as did nest in it? It surely attracted more kinds of resident birds than the average house. Perhaps its unusual architectural features, its isolation, and its unique location in a field surrounded by mixed forest all combined to endow it with special allure. For us human beings it possessed an elusive rustic, woodsy aura, and I often think that it must project something special to each avian perception. Perhaps each bird "sees" in it something particular that stirs its specific ancestral memory. And why not? Each of us in the family sees it differently, and chance remarks from guests reveal other perceptions that might not have occurred to us. In a sense, the bungalow is to each of us, as probably to each bird, what each "thinks" it is.

9 ○ BLACK AND VELVET
IN THE ROUND

Low fragile and transient is tranquility! As if confirmation of my thoughts were required, a succession of angry howls from the direction of the back door shattered these reflections as well as an otherwise serene morning.

The serenity of the morning, so precious to me at that moment, was a bane to others. Earlier while trying to call the local lumber mill on our party line I had broken in on the conversation of two neighbors, one of whom was remarking sadly that according to the Coast Guard report there would be no wind for sailing until mid-afternoon. This was one of those days which in past decades frustrated the skippers of coastal schooners. On just such a morning a schooner loaded with lumber in East Bluehill harbor would have had to be towed painfully slowly beyond the headland by a crewman bending the oars of a dory.

Another to whom the calm came as a calamity was a male loon that wished to leave the bay, for some reason known only to itself. Loons, superb divers, are abominably inept fliers. In order to gain air speed and enough lift, they require either an enormously long runway or a compensatory strong head wind. It is said that loons are occasionally trapped on small ponds in deep woods until a stiff enough wind invades the forest to provide them with lift.

This morning I had a telling demonstration of the constraints that they can suffer. The loon with wanderlust began its take-off from the glassy water near the sloop's mooring. With wings beating at maximum power and feet running at top speed along the surface the loon took off in the direction of Darling Island. Not until he was abreast of the island, a distance of one-quarter mile, did he break the surface. At that point his altitude was a matter of inches. One and one-half miles farther on—that is, just past the red nun marking Darling Ledge—he had gained about one foot in altitude. Now he began a slow turn to the east, passing Long Island, turning toward Newbury Neck. Still too low to clear the trees after four miles of flight, he continued his wide turn to the north, then westerly back toward the bungalow over which he passed at an altitude of about one hundred feet on an inland bearing.

It was this tranquility and communion with nature that Paul and Jehan shattered irretrievably with their howls.

Before I could get to the source of the disturbance the two of them, with arms flailing the air, came roaring around the corner of the house.

"What's going on?" I asked. "You've probably awakened all the neighbors and the whole village."

"Bumblebees!"

"Bumblebees on the back porch?"

"There's a nest there," they explained.

"Did you get stung?" I asked, rather unnecessarily.

"I did," replied Paul.

"I told him to run," Jehan added.

"Let's investigate," I suggested.

Invariably when we departed at the end of summer we forgot one or two things. The previous fall it had been the old army fatigues that I usually donned when venturing under the house to shore up the foundation stones. I had left the fatigues hanging from a nail in the inside corner of the open porch. The bees were in the fatigues.

By the time we arrived they had quieted down enough to allow us to approach cautiously. We decided, nevertheless, that some defense would be prudent should they again become belligerent. I dispatched Paul to find the insect net. Jehan had already armed himself with a can of insecticide.

Our interest in the nest was threefold. First, I would still need the fatigues on occasion. Second, a nest of potentially aggressive insects near the door posed some hazards. Who knows the subtleties of a bee's temperament? Third, I had always wanted to see the honeypot that bumblebees construct in the middle of the nest. (Also, I admit, I had a hankering to sample its contents.) It was not our intention to destroy the colony unless absolutely necessary.

Few insects have the emotional appeal of so many birds and mammals. Some excite wonder, as, for example, darting dragonflies; butterflies are admired, and slain, for their gorgeous colors; crickets are enjoyed for their singing but not loved as individuals on close acquaintance. But bumblebees are different. Bumblebees are fat, round, and fuzzy. There is something appealing about rotundity. To be round is to be jovial. Santa Claus is round. Falstaff was round. Friar Tuck was round. When you clad rotundity in neat black and yellow velvet the appeal is magnified.

Who has watched a fat queen bumblebee in the spring, freshly awakened from her long winter nap in litter in the field or in subterranean burrows, and not rejoiced in this harbinger of summer? She cruises in slow dignity close to the ground, examining decaying stumps, cavities in the litter, crevices among the grass clumps, the burrows of mice and other likely holes, searching for the ideal site in which to establish a colony.

She is conspicuous, she is secure, she is docile. Indeed, some of these queens can be handled, albeit gently. Instead of stinging immediately, they will struggle to escape, will fend off pokes with their legs, will bite. This docility continues to mark their demeanor later in life except when the nest is disturbed. If foraging, the bumblebees wish only to be left alone. In the garden they attend to their own business.

In her nest a bumblebee is something else again. She does not display the aggressiveness and ferocity of the yellow jacket, but she and her workers do sting when provoked. The bumblebee colony lacks the fighting hordes of the honeybee hive, but the small number is compensated for in part by the ability of bumblebees to sting repeatedly. Unlike the honeybee, which gives her life when she stings because she leaves the sting and part of her abdomen in the intruder, the bumblebee withdraws her sting after injecting the venom. She lives not only to fight another day but to fight again in the same battle.

All these things we knew as we respectfully approached the site of the nest. I also knew from past experience, as did Paul from his very recent experience, that a bumblebee stings with authority. To begin with, therefore, we watched activity from a safe distance.

There appeared to be about six workers flying circles around the fatigues. They seemed less interested in assaulting us than they did in patrolling the immediate environs of the nest. In less than five minutes, however, all disappeared into the folds of the garment. Two foragers arrived from the field, unaware of untoward events, and entered the region of one of the pockets. One, or a nestmate, then emerged to go nectar or pollen collecting. It was surprising how rapidly peace returned.

What should our strategy be? The nest could not remain in its present location. That was out of the question. On the other hand, we had no desire to annihilate these useful and harmless insects. Without the wild bees, the solitary bees, and the bumblebees, many a flower would pass its life unpollinated. Few would set seed. Honeybees were unable to fulfill the task alone.

Of course, one of us could rush in, snatch the fatigues, and so clear the porch, but that would not solve the problem of clearing the fatigues. Jehan volunteered to catch as many workers as he could with the insect net. His idea was to empty the nest so that it could be transferred to a new location. He tried. Confusion ensued.

"Run," he shouted.

"Ouch," I added as a worker found the end of a finger.

"Don't just stand there," Paul said.

As before, Jehan admonished, "I told you to run."

So much for curiosity.

I concluded reluctantly that we would have to destroy this colony.

At that Jehan advanced in a cloud of insecticide from the spray can. "There goes," I thought, "any chance of robbing the honeypot." It would be thoroughly contaminated.

In a matter of minutes all activity and sound ceased. Cautiously I began to search the fatigues. Finally in one of the pockets I found a wood mouse's nest. Within that was the small bumblebee colony. The spray had performed as advertised, so I could bring the mouse's nest into the sunlight for examination.

Viewing the remains at close hand I lost my feelings of guilt. The colony had been a small one. Inside the nest were a queen and six small workers. All except two workers had been killed by the insecticide, as evidenced by the flexibility of their legs and wings. The two workers, stiff and dry, had obviously been dead for some time. There were no combs and no honeypot. This had been a sick and dying colony. The reason was evident. The nest had been invaded and infested by the greater wax moth.

Honeybee hives are sometimes invaded by the lesser wax moth, a species whose larvae feed on wax and debris of the combs. Here we found the caterpillars of a large wax moth, an insect native to Europe and uncommon in the New World. The bumblebee nest swarmed with the naked, mealy-colored caterpillars that had spun masses of very tough silk webbing throughout the old mouse's nest and everywhere in the hollowed interior where the bee's nest had been.

Here we saw an example of subletting twice removed. A wood mouse had built a nest in my army fatigues; the bumblebee queen had moved into the mouse's nest to establish her colony; the wax moth had moved into the bee's nest and established a caterpillar colony. No space is wasted in nature. No territory is inviolate.

Two curious features characterized this particular bumblebee colony. First, it had been parasitized by a rare usurper. Second, it was located in an unusual place. The queen, easily identified by her yellow thorax and first two abdominal segments and by the shagginess of her pile, was the Shaggy Black-tailed Bumblebee, and that species characteristically nests in holes in the ground. Some other species are surface nesters, and some occasionally nest in sheds and stone walls, but not this one. At least so the reports go.

Later in the summer we discovered by the same painful method another colony of this species in the woodshed behind the bungalow. This time, by working at night, we were able to remove the colony, also in a mouse's nest, and place it in a box provided with a glass cover and an entrance hole.

Although some colonies thrive in artificial nests, this one fared poorly, partly, no doubt, because of my inexperience in these matters. For a while we could watch the activities of the stay-at-home queen and her small but energetic workers. We saw the wax honeypot near the entrance. Returning foragers attended to its filling with nectar. Before venturing forth on trips they frequently refueled, nectar being the energy source that kept the flight muscles operating. By the time summer was half over the colony had produced several males and three queens. The nest was abandoned late in August.

Once more a cycle that had endured for millions of years had been repeated. The queen bumblebee in the woodshed had slept through the long Maine winter, had found herself a nest, had built combs, had laid eggs, and had foraged far and wide to bring nourishment to the developing workers. She had helped these out of their cocoons. From then on these small replicas of the queen had literally devoted their lives to providing for her and the larvae that were to become additional workers, males, and new queens. The young queens would eventually leave the nest and just before the advent of winter seek shelter in which to hibernate. Those that were not discovered by mice, eaten by skunks, or killed by cold or infection would go house hunting in the spring as had their mothers before them.

I did not leave my fatigues hanging on the nail again, but the bungalow offered many other tempting sites. There were also, I am sure, several cozy mouse nests here and there about the building that had escaped my eye but could easily be found by a meticulous and persistent queen honeybee.

We were perfectly willing to share our garden with these neat, rotund, fuzzy tipplers if only they would be more circumspect in their choice of housing. A garden with bumblebees is a complete garden. Beautiful as flowers may be, they are captive things, static without a breeze to play upon their resilience. Bees and butterflies give motion to a garden and with it a sense of being truly alive.

Plants, to be sure, do move, but the tempo of their growth is of a slowness that deceives the eye. For us motion must be neither too slow nor too fast. We cannot see the plant grow; we cannot see the beating of the bee's wings; but we can see the bee move. With the bee's movement our eye is led from one blossom to another, from one color to another, from one point of view to another. There is no need for a painter's artifice to mime the perspective, to lead, to emphasize.

Not by motion alone does the bumblebee animate the garden. She holds our attention by character and the air of purpose of her flight. At the same time there is a ludicrous yet touching mien to her activity. She does not seem designed for flight, and indeed a few decades ago an aeronautical engineer proved mathematically that bumblebees could not fly!

In England these portly insects are called humblebees after the Old English meaning "to buzz." The word "bumble" also means to buzz but equally to blunder. These bees do make us smile because they are so droll as they bump around among the flowers. Perhaps we smile because, as the entomologist William Morton Wheeler once pointed out, "they are clad in habiliments better suited to a more indolent and Sybaritic existence" and "we are so used to finding industry and bustle in our fellow men most frequently associated with a lithe and lightly clad physique."

Even after the sun had set we often watched from our screened porch that final unequivocal sign that the day had ended—the last, laggard bumblebee blundering in the fading light into the blossoms of the foxglove for a nightcap.

10 ∘ JAPANESE LANTERNS
AND CHANDELIERS

FOG! The marine forecast called for it to burn off by mid-morning. Until then, what a fine opportunity there was to write! There would be no interruptions, no temptations to do something out-of-doors, and no distracting sounds, because the fog muffled all.

I sat pondering before the fire and in the process stared into the rafters above as though inspiration were to be found there. It was not inspiration, however, that lurked above; it was evidence of yet another dweller in the bungalow. High up on the roof, near the ridgepole, I caught sight of a small, pale, paper-wasp's nest.

Here I sat writing on paper, and up there nearly hidden in the shadows hung the handiwork of a lineal descendant of the inventor of paper. It is said that the Chinese invented paper sometime around the first century B.C. And indeed it is true that before that time mankind wrote on clay tablets and parchment and papyrus. Nevertheless, paper as such was first manufactured several million years earlier.

Even when mankind finally got around to making paper, a wasp had a role in it. Until the eighteenth century manmade paper was manufactured from rice stalks, flax, or rags. The idea of using wood pulp came to the French scientist Réaumur as he watched a wasp constructing a paper nest.

Nests of paper-making wasps are not uncommon on the outsides of buildings, but they are not often found in living rooms. The nest that had caught my eye belonged to a bald-faced hornet, a rather large, formidable, aggressive member of the tribe clothed in funereal black and white, with a long face to match. To observe the activities of this particular hornet would, I reasoned, be hard on the neck. How much better it would be to see whether or not there were other nests more conveniently located for observation.

As I found out later that morning there were indeed three other such nests, two under the eaves of the house proper and one under the eaves of the screened porch. This one could be observed quite comfortably from a reclining position in the hammock, an observation post that earned me many friendly jibes from the family concerning my laziness.

The nest on the porch was in a very early stage of construction so that over the ensuing days I was able to observe a very ancient art of paper making. The artisan, a queen that had over-wintered in some secure crevice, spent more hours at her work than I was able to devote to observaton. I saw enough, however, to appreciate fully her techniques.

She was manufacturing a coarse, gray, wood-pulp paper. One of our well-weathered cedar posts supplied all the pulp she needed. On one trip after another she gnawed industriously on the weathered bark and wood, gathering mouthfuls of fiber which she chewed into a pulp mixed with saliva. She worked this gray mass very much as the mud daubers worked mud, but her material permitted her to produce much thinner layers than was possible with mud.

When I began my observations, the support of the nest and part of the comb had already been finished, but the work in progress illustrated beautifully how the paper was made from the pulp with the simplest of tools—mandibles and maxillae. The nest consisted of a small comb of seven cells attached by a slender but very tough pedicel to rough board. Beginning where the pedicel was attached, the wasp had started to fashion an envelope, such that upon completion the nest would resemble a small Japanese lantern. Straddling the edge of the envelope she applied a little bit of pulp at a time, patting it into a thin layer as she moved. It dried very rapidly. Close examination of that area of the envelope that was complete revealed the concentric rings as pulp of different textures of gray were spun, as it were, around and around.

With each added layer the envelope became even more lantern-shaped until the point when an entrance hole had to be left in the bottom. Here the hornet did not stop when the hole had been closed to the correct size. Instead she began to extend the envelope into a long tube. When the tube attained a length of nearly three inches, construction ceased.

Now the process of feeding the next generation began. With the small wasps inhabiting the chimney and the cedar posts the best strategy for studying feeding had been to watch individual wasps as they arrived at the nest. They had been much too small for me to scurry after in the field, so I simply watched them stuffing food into the nest. This was possible only because these wasps were mass provisioners. The bald-faced hornet by contrast was a progressive provisioner. She did not pack her nest once and for all and resume a carefree existence. She, like bees, and human beings for that matter, fed her offspring daily. Unlike bees she did not feed them sweets; she fed them meat, meat chewed and masticated to a pulp. She herself was a vegetarian; she fed on the sweets. There was little to be seen, therefore, by stationing myself at the nest. The strategy this time was to observe the queen in the field. With that in mind I daubed a spot of bright red paint on the queen's thorax while she was resting on the nest.

Contrary to what one might imagine it was not at all difficult to obtain frequent glimpses of bald-faced hornets as they pursued their daily activities away from the house. The setting of the bungalow in a clearing enclosed on three sides by forest and fronted by the sea provided, as it were, a self-contained world. Only larger animals and strong fliers habitually ventured farther afield. Smaller creatures tended to remain with us. The bald-faced hornets were among those that satisfied all their needs without leaving the immediate neighborhood.

During the weeks that followed I frequently observed hornets in the garden, the field, the back yard, and the shore. As often as not one of these would be the marked queen. She apparently gathered most of her nectar from the garden and from wildflowers in unmowed areas of the field.

One day as I was rescreening a door I observed her hunting. The door lay on two sawhorses placed near the woodshed which doubled as a tool shed. A warming sun shone on my back and lighted up every small detail of the shingled shed. I had stopped work for a moment because a half dozen or so stable flies had become too persistently annoying. These flies, often mistaken for houseflies, are possessed of stiletto mouthparts that give one an impression of having been stung whenever the fly draws blood. When they were not leeching me, they were resting on the warm shingles.

After having slapped at one and missed it, I saw it land on the shed. As I was debating whether it was worthwhile to get a can of spray, a bald-faced hornet struck like a hawk pouncing on a mouse. The fly had been too quick for me, but it more than met its match in the hornet. Hardly pausing to shift position she began to remove the wings and legs. At some point she would chew the torso to a pap that could be digested by her larvae. She was hunter, butcher, and nurse.

Like any predator, the hornet sometimes misses. Whether this is due to poor eyesight or miscalculation is not known. Some light was shed on the matter a few minutes after the first kill. Another provided evidence that hunting is done by sight and that the hornet eye has certain defects. This queen dived at full speed onto a nail head showing black against the silvery shingle. She had been able to detect the fly-sized head of the nail but lacked the visual acuity, at least until it was too late to check her dive, to discriminate between nail and fly.

Other characteristics of the bald-faced hornet's foraging and feeding behavior were revealed to us gradually during our frequent picnic lunches on the shore. On pleasant days it was our custom to lunch on a rocky ledge that jutted from one end of the beach, the same ledge with the cut drill holes that the mud wasps occupied. From time to time various beggars hung about to cadge crumbs. The dog Gulliver was always there. Patrolling herring gulls would check back from time to time. A chipmunk dropped by now and again. Occasionally a trio of yellow jackets and a bald-faced hornet made appearances. The red queen visited twice.

Her visits provided a beautiful opportunity for us to learn about the relation between the hornets and the world as they perceived it. One day we had steamed a few clams. A hornet arrived and landed directly on one of the empty shells. Almost immediately she found the stump of the muscle that had held the shells together. Oblivious of my efforts to shoo her away she attacked that very tough muscle. In a surprisingly short time she had hacked out a chunk of meat with which she departed. In the meantime two yellow jackets had discovered one of the other shells, and similar butchering took place.

It seemed as though the hornets had found the clams by orienting to the odor. To test this idea I moved the shells to another spot ten feet away. At first the hornets returned to the original location, which they had obviously memorized; but they wasted little time there. Within a few minutes they found the displaced shells.

Behavior changed at dessert time, the shells having been disposed of in the meantime. Dessert was blueberry pie. The hornets relished it. They were persistent. They even fought for places on the pie. But now they were not toting food to the larvae; they were eating their own meal. This time when I moved the dish to another place on the rocks it remained unattended, while the hornets searched persistently at the spot where the pie had been. They had located the clam shells by following odor trails; they apparently had found the pie by general exploring, but, having memorized the spot, were unable immediately to find its new location.

One other paper-making hornet also hung her nest from the eaves of the bungalow. Polistes, one of the most common hornets, is a long, thin, black and brown wasp with a spindle-shaped abdomen. Her nest consists of an open comb hung from a pedicel and resembling a gray chandelier. Since it lacks concealing paper envelopes, there is no privacy.

The nest begun by a queen that had over-wintered consisted at first of as few as five cells. In each of these the queen had laid an egg, and, as the larvae developed, she fed them a steady diet of macerated meat. From these larvae would come more workers and queens, all of which would be concerned with enlarging the comb, feeding the young, and guarding the colony.

On one occasion I had the good fortune of observing a Polistes on a successful hunt. She had been flying from leaf to leaf in our Astrachan apple tree where she appeared to be searching for something. What she found was a green caterpillar about her own size. Seizing it around the middle with her mandibles she pierced the tough body wall. Since caterpillars are kept rigid by the pressure of their body fluids, the wound resulted in sudden collapse. At this point the hornet began to cut her victim in half. This accomplished, she held one half in her front legs and started eating it. The seconds passed. Slowly the morsel disappeared until I began to wonder if her stomach were a bottomless pit. Eventually, however, she reached the point where it appeared that the final chunk just would not fit.

With the chunk of caterpillar meat grasped firmly in her widely spread jaws, Polistes methodically cleaned her antennae and all six legs. Only then did she fly to her nest. There she began to stuff masticated caterpillar meat into the mouths of the developing larvae. She never returned to the remaining half of the caterpillar. It was just as well. There was little to return to because three gaily colored flies were making short work of the carcass. They were the jackals of the insect world coming to the abandoned kill.

When Polistes was not away hunting, the nest was seldom left unguarded. If one of us poked the hornet, she raised her wings in a threatening gesture, turned to face the source of annoyance, and tried to fend it off with her legs. If we persisted, she took flight in order to attack. She behaved similarly toward a strange hornet of her own or other species. A sister queen or sister worker gave a proper password probably in the form of a recognizable scent, and was accepted.

Polistes gathered most of their sweets from wildflowers. They were especially fond of goldenrod and remained active until the last flower had withered, and the first frost had numbed them. Even then a warm sunny day would revive them for brief periods of exercise. Quite a few queens would hibernate in the bungalow itself.

The old queens would die, but the nests that they had hung in the eaves, protected from the elements, would last for years. It is ironic that the creatures that invented paper were never destined to write upon it. The only evidences of their having lived were paper houses, and these were not even recognized by future generations.

11 ∘ RUSTY RAPSCALLION

HANGING on the front porch is a ship's bell rung to summon anyone of us from the field, woods, or shore. Paul and I were just preparing to fell a moribund birch when the bell rang. As often as not its peals meant a phone call for me, but we both acknowledged the summons and returned to the house.

Lois met us at the door with a broom in her hands. "There is a squirrel chewing the rug," she announced. "I've chased it away twice already."

She led the way to the living room. There to her understandable irritation and our astonishment was the persistent squirrel highlighted in a patch of morning sunlight and struggling with the fringe of the rug. With feet braced he or she was tugging with the determination of a dog worrying a bone. As Lois went after her (subsequent events proved it to be a female) with the broom, the squirrel practically chattered defiance before running, not, however, without a mouthful of fringe, to the cedar column that supports the main beam. Up she scurried into the shadows of the rafters.

Now we faced a challenging problem. A squirrel in the house is potentially far too destructive to be tolerated. The immediate problem was how to evict her. After that we would have to discover where she had gotten in and close that entrance.

All of us were familiar with red squirrels. They excited our reluctant admiration, reluctant because we could not quite forgive their occasional appetite for birds' eggs and nestlings. Yet they are handsome, bold, saucy, and charmingly disrespectful. One of the joyous sounds of a Maine morning is the chatter of a red squirrel. They seem to possess much more character than the grays, which remind me of conservative old gentlemen, spry but proper. The reds are the cockneys of the tribe, vocal, argumentative, independent, and disdainful of human overtures of friendship.

Why one should elect to join us was not immediately obvious. I can understand the house's attraction for other inhabitants. For the little brown bat whose natural breeding and roosting places are caves and crevices the bungalow is simply a wooden cave and a collection of snug, dry, crevices. It has the advantage of being set in the middle of a large unobstructed space where flying insects abound. For swallows, swifts, and phoebes it is a cliff, a series of secure ledges, a variant of the solid, craggy, vertical surfaces provided by New England geology. It is even a logical abode for flying squirrels accustomed as they are to roomy hollow trees, dark places, and a nocturnal way of life. Smaller creatures than all of these presumably are guided by other features. With perspectives cast on a less grand scale they must perceive one part of the house or another merely as textured surfaces, geometrically appropriate contours, wood, stone, or other natural substances indistinguishable from any other wood or stone. Or they sense particular areas or pockets of congenial shade or light. Perhaps they are trapped, as it were, by ranges of humidity or temperature similar to those that their ancestors became adapted to eons past.

But what is the bungalow to a squirrel? Could anyone watching the rusty-coated rascal in his natural habitat predict that he might find anything about a house to his liking? I think of all the times that I have watched red squirrels in the forest. These are not the urbanized beggars of our city parks. Squirrel and forest! Each is the spirit of the other.

On those joyous mornings when nature is at its best I can hear a squirrel *churring* in the woods, his call nearly as rapid as that of the cicada. I can go with the stealth of an Indian along one of the several paths we keep open. I can, often, arrive unobserved at the foot of the spired white or red spruce from which the squirrel calls. Unaware of my presence he frisks to the topmost branches where the green cones bunch, picks off a cone, and descends to a horizontal branch where he sits on his haunches and proceeds to shell the cone for its seeds. A small shower of green scales filters down upon me. If I make my presence known, the squirrel scolds, angrily, in a staccato chatter. At times his call is scarcely distinguishable from the alarm call of a forest bird.

Not infrequently there would be four or five squirrels scolding from different sectons of the forest. One would be out on the headland where the red spruces dominated the community of trees. Another would be on the landward side of the road where white spruces clustered. One bold or foolhardy fellow would be heard from the trees along the brook where the broad-winged hawk had her nest.

On other days there would be no sound of squirrels. The forest would even be silent of birds. If I went to spots that I knew squirrels frequented, I often found a favorite moss-covered boulder with piles of freshly picked cone scales heaped upon it. It took little imagination to picture a red squirrel perched there, peeling a cone that he had fetched from a spruce top. I would wonder why he had come all the way to the ground to dine when obviously there would be long climbs between cones. But then what is such a climb to a squirrel that can ascend a tree faster than I can run on level ground?

Some days Gulliver would take it into his head to go squirrel hunting. For him the venture must have been pure sport, unadulterated pleasure, because not by the wildest stretch of the imagination, neither mine nor his, could the hunt ever have been rewarded with bounty.

The sport revealed as much of the nature of the red squirrel as of the dog. Gulliver seemed to hunt by sound and sight. He would wander through the woods, nose to the ground, occasionally lifting his head to test the wind, and frequently stopping to peer into the forest canopy. Unless a squirrel was already scolding, Gulliver initially detected him by the sound of movement in the trees. Thus alerted he would peer upwards. If the squirrel moved, Gulliver would track him by sight, all the while barking frantically. Usually the squirrel sat on a branch whence he scolded back. He would chatter, he would chirp, he would cluck. As he clucked, his tail jerked in such violent accompaniment that one almost feared for the security of its attachment.

The squirrel apparently considered the hunt fair until I arrived at the scene. When that happened, the squeaks, chattering, and churring ceased. Quietly he would run to the top of the spruce, and, if it was thick enough, would hide silently. Or, he would run softly from tree to tree until he had put what he considered safe distance between himself and us. Sometimes Gulliver would see him going and in full cry would follow along the ground. If the squirrel succeeded in giving him the slip, he would mount guard beneath the last tree in which he had seen the quarry.

This then was the natural home of the red squirrel, the lofty spruces and firs where cones were abundant and concealment ever possible. In these scented galleries, he played, courted, and feasted, coming to the ground occasionally, where he sought mushrooms, a special delicacy, but generally leaving the boulders and burrows to the chipmunk.

In conifers he also nested. Against the winter cold, if he could not find a knothole or abandoned woodpecker hole to his liking, he built a substantial, lined nest, as tightly woven as a basket, warm and secure against the elements. This also became the family home. For the warmth of summer, he had the equivalent of our summer bungalow, a lighter, more airy nest.

We employ the same word, "nest," for the elaborate structures built by birds, wasps, mice, and squirrels. Properly speaking, the nests of wasps and birds are cradles. They are receptacles in which the young are born (even though birth may be as an egg), nourished, and cared for until they are able to fend for themselves. The nests of mice and squirrels also serve this purpose, but over and above their role as cradles they are larders for the mice who store nuts, seeds, and berries; and they are true domiciles.

The red squirrel keeps nest and larder separate. He stores food in special caches under or on top of the ground. Stores are generally restricted to cones. Many items suffice as daily fare in the spring and summer. In the early spring he eats buds, catkins, flowers, and maple sap, which he obtains—if there are sugar maples in his territory—by slashing the bark and lapping up the exuding liquid. As the season progresses he eats fleshy fruits, seeds, insects, birds' eggs, fledglings, and mice. He will even eat young cottontails and any bird that he can catch.

His favorite food, however, is the cones of spruces and firs. What he does not eat on the spot he cuts and lets fall to the ground. After he has cut ten, twenty, or even more, he descends to gather them into a pile at the base of a log or in an underground cache. Caches may contain bushels of cones, and the same cache may be used year after year.

What, then, attracted the red squirrel to the bungalow? She hardly seemed like the domestic sort. How could such a cheeky free spirit reconcile herself to the confines of the house, even if it did offer some of the solid snugness of a knothole? There certainly was no food available. We saw to that as a precaution against giving the wood mice a sense of hospitality that we did not wholeheartedly feel.

The very idea of a squirrel nesting in the house overcame my usual habit of procrastinating in matters of duty. It was clear that I must: get the squirrel out of the house before she discovered that overstuffed chairs offered even softer nest material than rugs, find and remove the nest if there was one, and block whatever hole had afforded entry in the first place. None of these was an easy task. Chasing a squirrel around the inside of a bungalow like ours could be infinitely more difficult than the frustrating business of trying to catch one small, humming mosquito in the bedroom at night.

Lois had last seen the squirrel disappearing in the maze of beams and rafters overhead. Our first step obviously was to search the upper recesses from one end of the house to the other, a tricky business because these upper regions had all of the charm and surprises of an attic. Shadowy, inaccessible, abundantly supplied with cul-de-sacs and cobwebs where a whole colony of squirrels could hide, this pseudo-attic also presented the challenge of avoiding the occasional shingle nail poking through the roof.

At this point, we had not even formulated a sensible plan for removing the squirrel if we found her. The mind boggled at visualizing the chaos of a chase in an area which to the squirrel would be simply another tree with squared instead of rounded limbs. Nevertheless, seek we must. We brought a ladder and a powerful flashlight and began with the two storage lofts. Numerous interesting discoveries were made during the search: a hand-cranked ice-cream freezer, old camping equipment, an ancient tennis racket, two hurricane lamps, the usual trunks, an old highchair, several mouse nests, but no squirrel. Methodically we worked our way toward the center of the house. Still we found no signs of a squirrel, nor, for that matter, of any conceivable points of entry. We saw places where a small brown bat could squeeze in but nothing that would accommodate a red squirrel.

Next came the region of the chimney. The rough granite fireplace was massive. Above the mantle, the stonework narrowed abruptly forming a wide ledge tucked against the rafters where the chimney made its exit through the roof. When I flashed the light into this recess, four small lights shone brightly back at me. Moving the light beam I made out the forms of two young squirrels whose rusty coats blended almost perfectly into the brown background of timbers. The two crouched in a pile of soft material in which damning evidence in the form of pieces of rug fringe was clearly visible.

For several seconds we stared at each other. At least I stared at them; they stared at the light. The two were nearly full grown, quite capable of caring for themselves. Appealing as squirrels are, these two were exceptionally attractive in this setting. Snubby-nosed little vandals, I thought.

My next reaction was perplexity. Red squirrels raise two litters a year. In this part of Maine the first young are born in the springtime, the second litter during the first part of the summer. I guessed that the young were part of the first litter. Why then were they still occupying the nest, if indeed this was the nest of their birth, and why was the mother, if it had been she, tearing material from our rug?

Answers to these questions would have to wait. Turning to practical matters I decided on a plan of capture. At my request one of the boys brought a pair of heavy leather work gloves. Since there was neither room nor time to resort to nets or traps, I reached behind the chimney, one hand at a time, and grabbed the squirrels while they were still mesmerized by the light. Descending the ladder with a struggling, biting squirrel in each hand taxed my latent acrobatic skills; nevertheless, I managed to descend, run for the woods, and as a last humanitarian gesture release the pair at the foot of a balsam fir up which they scrambled.

Returning to the house I resumed the search for the mother and for the entrance hole. The latter turned out to be, in all probability, a squirrel-sized gap near the sill between the wall and chimney. At any rate no other opening was found. A piece of tin and some roofing nails corrected the defect.

We never did find the mother. She must have been outside while we were searching and foreclosing her home. We did not expect a return visit. As was the case with the bumblebees and the wood mice, the red squirrels had to be numbered among those tenants whose adoption of our house was flattering but unwelcome after the first encounter.

Of all our tenants the red squirrels were the most clever. They could not construct artistic paper lanterns nor engineer exquisitely precise webs nor sing glad or haunting songs. They could not navigate skillfully in the dark nor voyage to other lands, but they were clever. In the summertime they had little opportunity or need to display this gift, but their ingenuity in solving the puzzle of bird feeders in the wintertime is legend. Every time someone invents a squirrel-proof feeder the squirrels outwit the inventor. We have watched our squirrels trying to get to a long tubular feeder hanging from a wire midway between the house and a nearby tree. Usually the squirrels are content to feed on seeds spilled on the ground by the birds. One day, this supply being exhausted, a squirrel displayed his amazing problem-solving abilities. The ground now being barren of seeds, the squirrel sat under the feeder and stared upwards. Had he been a person, I would not have hesitated a second in declaring that he was sizing up the situation. Had the feeder been a steak and Gulliver the hungry animal, he would have exhausted himself leaping for it. The squirrel, on the other hand, seemed to have measured the distance and decided that it exceeded his range. Whether he was in fact "measuring" and "deciding" I of course have no way of knowing. But he appeared to be.

Next he resumed hunting on the ground but, finding nothing, again riveted his attention on the feeder. This time his gaze followed the wire to the tree. There was no doubt about it. In a matter of seconds he ran to the tree, scurried up to the wire, hesitated for a moment, during which he stared at the feeder, and then began his tightrope walk to success.

This was but one of many examples of problem solving. It can be argued, quite reasonably, that the situation with the feeder was just a particular form of a general problem that the squirrel encountered frequently in nature, like a cone just out of reach in the next tree, or out at the end of a thin branch. Therefore, he could have solved this genre of tasks earlier by trial-and-error learning, no very remarkable feat. On the other hand, if the feeder were a new puzzle, never before encountered, and solved at the first attempt, without trial and error, he would have been reasoning.

Reasoning is sometimes defined as the ability to solve complex problems with something more than trial and error or habit. Whatever we choose to call the squirrel's behavior, he is undeniably clever. When we encounter him or his mate in the bungalow, we know that we are dealing with an adversary worthy of our respect.

Ｏｎ ｅ August night while I was attempting to read a very dull book, a trivial incident opened my mind to the possibility that an astonishingly large tribe of very ancient creatures resided in our house. The incident was the descent from the darkness of a modestly marked, small, delicate spider. Paying out a nearly invisible line of silk attached to a rafter overhead she swung daintily onto the page before me. She herself was not ancient, nor were her relatives, but she could trace her lineage to a stock that was already old when insects were newcomers on earth. She reminded me by her presence that, in the best Maine tradition of generation after generation occupying the same house, we were host to generations of spiders that were born and bred, that courted, reproduced, and died all under one roof.

Much maligned because of the misbehavior of tarantulas, black widows, and the brown recluse, the majority of spiders are quiet harmless guests. True, many disfigure and dirty the house with unartistic cobwebs that accumulate dust and aerial debris. But they atone in a small way by capturing some of the annoying intrusive transients. They reduce the numbers of moths—clothes and other kinds—blackflies, no-see-ums, pill bugs, and even carpenter ants—in a token way.

Of all our animal guests, the house spiders are certainly the most unobtrusive. This difference makes for unexpected encounters, and it is precisely the element of surprise that causes people to harbor an antipathy on first acquaintance. If a person were meeting these spiders for the first time, innocent of culturally acquired aversions, ignorant of the existence of poisonous members of the group, unaware of their relationship to less aesthetically pleasing members of the tribe such as hairy-wolf spiders, what would he see? He would see first a small creature less than half an inch long—a delicate creature. Admittedly she is no beauty. There is something about the symmetry that does not conform to our idea of beauty, but it is not ugliness. To me the aesthetic flaw in the design is that the two ends of the spider are incongruous. The first two pair of legs, long and dainty, loom in front, giving the impression of aggression and attack, whereas they actually provide the spider with an extraordinary tactile sensitivity. With their length and sensitivity they serve the spider in much the same way that long tapering antennae serve an insect.

The disproportionately large globular abdomen seems out of harmony with the slenderness of the legs. I am reminded of certain excessively obese people whose delicate hands and ankles betray the pathology of their obesity. The incongruity is the disturbing feature. And with the spider, even though the proportions are normal, the usual appeal of jocund rotundity so apparent in a bumblebee is here lacking. Perhaps the naked shininess needs to be clothed in velvet.

If one can convince himself that beauty is skin deep and that these spiders are neither aggressive nor harmful to people, he can begin to admire them for their not inconsiderable talents as producers and spinners of silk. A web is the hallmark of most species. Were it not for cobwebs, for example, the presence of spiders indoors would probably not be known to the average householder.

Even the homes of the most meticulous housekeepers contain cobwebs. Brushed away today they reappear on the morrow. Yet few people have actually seen the architects of these webs. This contrast between universal familiarity with cobwebs and almost universal ignorance of their builders crossed my mind as I watched the small spider disappear between the spine and cover of the open book. I realized that I too had not made the acquaintance of many house spiders.

What better time than now to correct the omission. House spiders retire during the daytime but are very active in darkness. With a flashlight I could explore the lofts and other nooks and crannies that might have escaped the dust mop. Exploration would provide me with a census, but unless I chose to spend a large part of my life in the loft at night I might not be able to identify the species or learn anything about their habits. Accordingly, I carried with me a few shell vials with which to capture live specimens for later transfer to some old cracked aquaria.

Altogether I captured three different species of cobweb spiders, and I would not be surprised if more turned up on further search. The three were: the common American house spider, a light-colored mottled Theridion, and a dark Streatoda with a whitish T on the upperside of the abdomen. All belonged to the family of comb-footed spiders, so named because the last segment of the fourth leg is equipped with a fine comb of bristles which the spider uses to draw silk from her spinnerets and fling over her prey.

From the moment of leaving the egg, spiders are tethered for life. One of the first acts of a spiderling is to spin a dragline fastened by adhesive to the substratum. From that time onward, until death, she is on a leash of her own making. We tend to think of a leash as a restriction on freedom, a form of imprisonment. For the spider, paradoxical as it may seem, the dragline is the secret to freedom and security in a hostile world, freedom because the line is infinitely long, literally without end in that it is spun until death. The spider can go anywhere she chooses. She can spin the long line into the limitless sky until the wind picks it up and snatches her into the air. There she may go ballooning for hundreds of miles and at altitudes sometimes as great as ten thousand feet.

Draglines offer security because the spider can fasten them where she chooses. She can rappel to the ground to escape enemies; she is ever sure of her footing because there is always the dragline to check her fall in case of a misstep. Even small jumping spiders jump from an attached dragline so that they can make daring leaps into thin air with impunity. For other spiders the dragline is the structural unit of the web.

With my spiders separated in individual glass containers I could move them from dark corners where I kept them during the day to dim light at night when I wished to watch them at work.

A cobweb, I learned, is not a mess totally devoid of plan and form. It is a most primitive type of web that must have evolved early in spider history, but it makes sense, as I soon discovered. Each of my captives wasted little time in exploring her prison. Each constructed a typical cobweb, clean, white, unsullied with dust. There were species differences, but the overall pattern was the same. It consisted of a criss-cross scaffolding, a tangled space web the threads of which outlined triangles, rectangles, trapezoids, polygons—almost every geometric figure imaginable, smaller and denser near the middle, looser toward the periphery. Some of the threads were beaded with sticky drops as were the guy lines attaching the maze to the floor and corners of the aquarium.

By tossing mosquitoes into these webs I learned how the snare worked. When my aim was accurate so that the mosquito landed in the maze or on a sticky guy or trap line, the spider rushed from the corner of the web where she had been resting. Manipulating the silk from the spinnerets with her comb feet she threw sticky threads around the mosquito. Unlike the orb spinners which wrap the prey by rotating it, she moved around the mosquito as she wrapped it. Only after it was hauled to the corner retreat did she bite it. With minor species-characteristic variations this was the mode of capture employed by each of the three species living in the bungalow.

One evening I discovered that Theridion had produced an egg sac. This was a pear-shaped, whitish sac containing brown eggs. About three weeks later approximately thirty spiderlings appeared in the web. Unlike many species in which the young leave to fend for themselves, Theridion spiderlings live with their mother sharing food in the maternal web.

I kept the family in captivity hoping to observe a remarkable parent-offspring behavior that occurs in one British species. There is in Britain a Theridion mother that feeds her spiderlings from pre-digested food. Each young spider slides down its dragline to the mother's mouth to be fed. I did not observe this touching sight in my captive Theridion; nor do I know whether feeding is uncharacteristic of the American species or whether I missed feeding time.

Cobwebs, representing the tangle-maze model of webs, were probably the prototypes of the more sophisticated models spun by those spiders that are more familiar to the average person. A first step in modernization appears to have been the addition of a sheet in the middle of the maze. The result was the production of some of the most delicately beautiful webs to be seen. It is hardly surprising that the artists in silk, the sheet-web weavers of the family Linyphiidae, were too fastidious to reside in the bungalow. Then again, they preferred different patterns of light and shade than any building offered. Also, they needed the elaborate scaffolding that the complex arborizations of bushes provided. They depended for sustenance on more flying insects than the bungalow offered even at the height of the blackfly season.

In early morning when the angle of the sun was just right, I could observe two of the more innovative species by walking along the woodland path to the headland. Here where the sun shone only in patches, where the air was usually moist, I found webs of the filmy dome spider and the bowl and doily spider.

The filmy dome spiders were wont to spin their webs in bayberry, huckleberry, spirea, and viburnum bushes which offered many points of attachment for the threads of the maze. In the center of the maze, the spider would erect a lacy dome evocative of the dome of a ghostly basilica. Beneath the dome she waits. Mosquitoes, gnats, aphids, midges, and other small flies blunder into the maze and fall through to the dome. From inside, the spider bites and pulls the insect through, shrouds the victim in silk, and after a while repairs the dome.

Among the same bushes the bowl and doily spider constructs her version of the maze-and-sheet web. In place of a dome she spins a shallow upturned bowl. Beneath the bowl she adds a sheet, the doily. This web operates in the same fashion as the dome web. In both webs the sheet, whether it be dome or bowl, protects the spider from predators from above. The doily is an added protection from predators that might approach from below.

Back in the house I found two other kinds of spiders, but after what I had seen they held no appeal whatsoever, neither in themselves nor in their webs. Both belonged to the family of funnel weavers. One was a common grass spider, Agelenopsis by name, not usually found in houses; the other, the European house spider, Tegenaria, an immigrant that probably arrived with the earliest settlers. The two are hunting spiders that have not given up the use of silk for a dwelling place. They are lean, long-legged, strong spiders, and extraordinary runners. These combinations of characters endear them to no one. It is they who evoke fear and loathing especially when they run at top speed across the floor.

Even though I was spider hunting in the bungalow, discovery of the grass spider startled me. She had constructed a funnel nest in one corner of a window behind a drape. The funnel, open at both ends so that retreat through the back door is possible when danger threatens the front, is normally placed under stones on tussocks or grass or, as in this case, in an inaccessible corner. The front yard, so to speak, is a broad expanse of dry sheet web which is added to as the spider grows older. It eventually becomes extensive and thick. Sheets like this are the cobwebs (quite a different thing from the cobwebs in a house) that are seen covering lawns when a heavy dew has fallen.

The spider sits in her funnel waiting for some insect to blunder onto the silken front porch. So fleet of foot is she that she can outrace any unfortunate cricket, grasshopper, June bug, or other trespassing insect. Before they can move from the sheet, she is upon them, delivers a lethal bite, and drags her meal into the funnel for leisurely consumption.

When I startled Agelenopsis behind the drape, she rushed out of the funnel, thereby startling me in turn. Finding no prey she retreated as swiftly. This time, without compunction and acting almost instinctively, I squashed the funnel, spider and all, into the corner. There would have been no opportunity to catch such an agile creature alive. Besides, for study I could observe grass spiders leisurely on a sunny lawn. My experience and reaction to Tagenaria was the same, and I evicted it as well.

Although we continued to maintain the bungalow in as tidy a condition as good housekeeping demanded, there was not the slightest chance that the spider population would ever be exterminated. This realization would disturb some people, and indeed the thought of living in a house with a limitless and self-replenishing population of spiders could be unsettling. We accepted this as a fact of rustic life and dutifully swept up cobwebs as they appeared. Every now and then, however, when I was at some sedentary occupation—like reading the dull book that launched me on my study of house spiders in the first place—and a spider appeared, I would watch in appreciation.

Even as I am writing this evening at my desk in the corner beside the fireplace, I notice a small spider running a line from the edge of the lampshade to a shelf of books. I do not know which species she is, and I shall not disturb her. She appears to be constructing a small web appropriate to her size. Perhaps before a new day breaks she will snare a blackfly or a no-see-um. For this I salute her. In the morning the desk will be dusted off, the web demolished. The spider herself will be safe in some crevice.

She will be Penelope. Her web will never be completed, because what she spins each night I shall undo each day. Nonetheless, by September she will have produced offspring, and after I am gone there will be spiders in the bungalow as long as there is a bungalow. And there is no good housekeeper who can gainsay that.

13 ∘ ORBS

ⓄNE of the wonderful aspects of nature is the manner in which it can generate aesthetic irregularity out of regularity. A contradiction, you say; but consider. Each detail, each unit, is designed with great precision and developed with ordered uniformity. Look, for example, at the needles of a white pine, five to a cluster, each 72° of arc; and each cluster is like every other cluster. Or, consider the red pine with three needles to a cluster, each 120° of arc. Look then at the whole tree. The two pines are characteristically different in form, but no two white pines are alike, nor are any two red pines.

A similar analysis could be made of the detailed form of any living plant or animal. The units are regular; the totality is irregular. Even you and I, bilaterally symmetrical as we are, are less than perfectly so, as has been so strikingly proven by composing photographs using one side of the face twice. Two different persons emerge from the all-right portrait and the all-left portrait.

Place all the irregularities of composite form together; compose them further as a field, a meadow, or a forest; mix the species, the grasses, sedges, and forbs, the pines, spruces, balsam firs, poplars, maples, and others; scatter a few boulders; roll the land into hills and vales; add a brook or two; allow the sun to mottle the composition with ever-changing light spots and shadows. Do all these things, and you produce an irregularity that is the quintessential glory of the natural. It is the beauty of the constantly unexpected, the constantly novel. How can a person ever be bored with nature? How can one ever adapt, in the sense of no longer being stimulated, to such complexity of form, light, and color? As every artist knows, no forest is the same from hour to hour or the sea from minute to minute; no day ever repeats itself.

It is this irregularity, this rich complexity, and this never ending unpredictable change into which animals must fit harmoniously. At the same time they must average these differences. Perceiving the variability they must never lose sight of the whole. Even though no two white pines are alike, they preserve the universal species characteristics while retaining their individuality. The nesting pewee must recognize pines as pines yet distinguish hers from others.

The artifacts of man, on the other hand, are characterized by regularity, uniformity, repetition. The bungalow as an example does indeed present a multitude of faces, of textures, forms, and micro-environments, but not without displaying a regularity nowhere to be found in nature. Every shingle on the roof is the same, laid row on row in perfectly straight lines. The shingles on the side, despite differences in width, run in even courses. The eaves extend the same distance on all sides; the rafters are identical and evenly spaced; all corners are plumb, or were when the building was young.

So, while the bungalow is many environments to many animals, it possesses repetitious features that can affect some of the animals living there. I have already alluded to the effect of evenly spaced rafters on robins, a repetitiousness that prompts some to attempt to build many nests. If one site is attractive, why should not the identical adjacent one be equally attractive and so on until all spaces between rafters are filled with nests? Furthermore, the profusion of attractive sites attracts more robins than would any tree so that only the bungalow can boast a surfeit of nests. Thus while the diversity of animals living in the bungalow can be attributed to the diversity of ecological niches, the profusion of individuals must be accounted

for in part by the redundancy of niches. Nowhere is this influence better illustrated than under the eaves.

I happened late one afternoon to glance up at the long eaves extending the length of the west side of the house. The angle of the sun favored the inspection. It glinted on an incredible row of orb webs, one web between every pair of rafters from one end of the house to the other, thirty-two webs in all. Where in the natural environment would one find so many orb webs so close together, so evenly spaced, and so evenly slanted that they reminded me of windmills all set to catch the same prevailing wind?

The machinelike regularity of the array imparted to it all the monotony characteristic of assembly-line productons; and the nearly precise duplication of pattern in web after web struck me as a needless overstatement. A more perfect union of methodical constructions, the carpenter's and the orb spinners' could hardly be imagined. Obviously the collaboration had not been planned. One cannot help wondering what pleasure man finds in precision, perfection of symmetry, and also what moves the spider to achieve the same ends. Perhaps only harmonious irregularity requires genius. Man has to strive mightily to prevent his art from appearing machinelike in execution. He must exercise his imagination to avoid banal reproduction. Or else, he must be lazy and inept to escape the same sterility.

The spider in a different way is bound by the conventions and strictures of her genes. The webs under the eaves were not absolutely identical because of strictures imposed by the house. To a certain degree the orb spinners' webs are the same whether they are hung from the iris in the garden, between two telephone lines, among the branches of a spirea, or across a path or brook. Each web differs in slight details, however, according to the space it is designed to sieve and the idiosyncracies of individual builders.

Much has been written about the orb spinners' techniques of web building, but to appreciate the wedding of engineering skill and artistry a person must observe for himself. This can be an absorbing pastime given the inclination and time as well as comfortable circumstances for observing. The spider readily cooperates by building a new web if her existing one is destroyed. The only restriction, at least for the orb spinners, is that spinning is a nightime occupation. My study of nocturnal activity was aided, however, by the construction of the bungalow. The side of the building with the webs was lined with a succession of eave-to-floor windows. After sunset the lights of the living room illuminated the spider and her silk threads against the black backdrop of night. Accordingly, I could relax in a comfortable chair, safe from blackflies and mosquitoes, and watch the entire performance as though in a theater.

Act 1, scene 1 could have been titled "The Bridge." The spider, having fastened a dragline to the corner to which she had retired during the day, walked along the underside of the roof and payed out a slack line. Arriving at the next projecting rafter and turning, she proceeded almost to the end. At this point she drew the line taut and anchored it. She then traveled back and forth along the bridge thus formed, reinforcing it with additional threads of heavier caliber. On the penultimate crossing she trailed a free line fastened at the point of departure and again at the destination. Now she made a final recrossing, this time on the sagging line. At the point of maximum sag, more or less the midpoint, she dropped a plumb line, with herself as plumb bob, all the way to the ground, in this instance the frond of a five-foot-high interrupted fern, one of several planted along the house. This line was pulled taut and fastened.

By these actions the spider had constructed a silk Y, the stem anchored on the fern and each arm on a rafter at the bridge ends. The intersecton of the arms of the Y was destined to become the center of the orb and the stem and arms themselves the primary radii.

In scene 2 the spider added more radii to the triangular spaces between the bridge and the two arms and between the arms and the stem. In the process other lines were added to the frame to complete the triangles. All these additions were accomplished by carrying free lines, anchored at the center, along existing radii and then walking a calculated distance along the tangential frame line to make the attachment. Several times during the placing and tightening of radii the spider returned to the hub to reinforce it with a mesh of silk. Such reinforcement of the focal point of an increasing number of converging lines, together with the addition of a few spirals, was obviously an engineering necessity.

Act 2 was brief but of great necessity. It consisted of spinning a spiral thread from the center (actually a short distance out from the central mesh and spiral), to the circumference. Advancing centrifugally the spider was eventually forced by the diverging radii to crawl to the outer end of one with a slack line, then back down the other to the desired point of attachment, there to tighten and affix the line.

The importance of this scaffolding line cannot be overestimated. As others had done before me, I went out on one occasion, destroyed a just-completed spiral, and noted what happened next. What happened next was a catastrophic outcome of instinctive behavior. Just as in an automated assembly line, each successive act followed in sequence regardless of whether the preceding act had been completed properly or at all. "The moving finger . . . having writ moves on." The spider began to lay down her final spirals. With nothing to hold the radii to their spacing and no scaffolding bridges to cross she trailed each long sticky thread out along a radius to its end and back in along another, leaving behind an excessively long sagging mess.

Act 3, the final one, built to a climax with the laying down of the sticky spiral that constituted the snare. The spinning reversed the order of things by beginning at the circumference where the radii had their widest separations and moving centripetally. Here is where the scaffolding proved its worth because the spider bridged the interradial gaps by transversing the dry spiral. As she went she spun out an elastic viscous thread. Crossing a gap she employed one leg to stretch the thread to its limit, attached it, and let go. With a snap it tightened across the gap. When the spider completed one circuit, and was attaching the next line of the spiral, she calculated the exact spacing of lines by measuring with her leg the distance to the previously laid thread. In the course of the spinning she cut loose the scaffolding which she then rolled into a ball and discarded.

The whole performance that I watched lasted about forty-five minutes. There would be no encore unless I again destroyed the production, a form of applause for which I had no heart. If I wished, I could return for the next evening's performance because the webs, sadly tattered at the end of the day, were sometimes completely rebuilt each evening.

Quite apart from its visual beauty and engineering marvelousness, the orb web is an exquisite vibratory instrument. Constructed on many strings of different lengths, under various degrees of tension, it is played upon by wind and rain, insects large and small, and courting males. The spider could be likened to a skilled musician who, if she could, would listen to a played piano by laying her hands upon the strings. So talented is the spider's sense of touch that from her position at one corner of the web's stays she can pinpoint the location of a struggling captured insect, judge its size, and distinguish it from the tune of the wind, the touch of a falling leaf, or the particular courting signals of an ambitious male. Just as she can assess the tautness of each line during the period of web construction, and know at which points additional guying was required, so can she judge by vibrations the location and character of the prey.

Upon receiving an appropriate signal she will plummet to the hub of the web on her dragline and unerringly run to the precise coordinates where the prey struggles. Still by touch she will immediately enshroud the fly or moth in a sarcophagus of silk. Unlike the comb-footed spiders that move around the victim as they spin, she spins the victim like a roast on a spit. She then administers a lethal bite, cuts the frightful package loose, attaches a line to it, and, ascending the dragline, carries the bundle to the hub or to her lair for dining at leisure.

As every competent fisherman knows, there is a time to fish and a time to mend nets. Many aerial web spinners repair soon after a catch is made. Some of the orb spinners make minor repairs even while bringing home the prey, cutting away a tangle here, adding a thread there. By and large the orb spiders under the eaves frequently spun new nets from the original bridges.

For them, as for all the spiders in the bungalow, the season of fishing had been a long one. At the beginning of summer when the first warm days of May had begun to drive the frost from the ground, soften the landscape, and provide sheltered pockets of warmth under the eaves, the eggs of the previous summer hatched. Several hundred spiderlings remained for a while as a living embroidery covering the formerly brown egg mass. Within a few days they scattered, each seeking a corner of her own. There she spun a minute but perfect web.

Soon there were more than a hundred small webs so delicate as to be nearly invisible. As the summer advanced, these became larger in size and fewer in number. Some spiderlings had ventured farther afield. Many fell prey to chronically hungry predators, even their own sisters or distant relatives. A spider's worst enemies are other spiders.

As the webs became larger, the best vantage points to accommodate them became fewer. Thus it was that by a process of having to fit larger webs into a limited number of available geometically suitable confines the spiders came in the end to fit uniform webs into the uniform spaces between eaves, and to produce only as many webs as there were spaces.

The few males that survived the dangers of growing up built rather pathetic looking webs wherever they could find space unacceptable to females. Beside the robust, talented females the males appeared small, puny, and inept. The females were large with eggs; but all the trapping of the preceding weeks, all the energy that had gone into egg development, would come to nothing if the male could not fulfill his destiny. To find fulfillment he had to risk his life.

With comb-footed spiders connubial bliss, or at least tolerance, was the rule. The mates got along well in a shared web. By contrast, the male of the orb spinner was a necessary but not particularly welcome partner. He had to convince the predatory female that he was not just another morsel of food. To insure a welcome reception he had to play a string serenade.

From the comfort of the living room, I was fortunate enough to witness one courtship from a distance of three feet—close enough to see the courtship but not its consummation. In order to present a complete account I must, therefore, integrate my observations with the reports of others.

Before embarking on a night of courting the male must prepare himself. To this end he builds a minute web, a few square millimeters in area. Upon it he deposits sperm. He then soaks his palps in the drop. The palp is so constructed that a sac therein becomes filled. Now his search for a mate begins.

Having found the web of a female, the male approaches cautiously and begins to pluck the strings. The vibratory signal is characteristic, unique, and instantly recognized by the female as that of a suitor. If she is not receptive she can rush the male or shake him loose, in which case he may escape by dropping on his dragline. But if she is open to his attentions, he is able to approach. How different sexual mores can be! With his palpi the male places the sperm in a special receptacle that the female has for storage. This bizarre ritual having been completed, he must escape immediately. To tarry is to die. Sometimes he escapes for another mating, sometimes he becomes a banquet. In the courtship that I observed he escaped.

By the time we were ready to leave for our winter home the orbs were in tatters. Egg masses had begun to appear in sheltered corners. The once-resplendent spinners were shriveled, shunken relics, dead or dying. Another cycle that had been repeating itself for millions of years had come full turn.

14 ∘ CRICKET ON THE HEARTH

ONE August afternoon my neighbor came through the woods to tell me that there was a cricket in her house.

"How fortunate you are," I remarked. "It must make a cheery sound."

She looked at me suspiciously to see if I was teasing. "He is driving me crazy. He chirps all afternoon and keeps me up at night with his racket."

"He makes a good burglar alarm," I said. "Did you know that the Japanese have kept crickets as watchdogs? When somebody enters the house, the cricket stops singing."

"Will you come and get rid of my cricket?"

I agreed to survey the situation, so together we returned to the house. Of course when we got there, the cricket stopped singing. Patience is an easy virtue on a lazy summer day so we waited and chatted. Sooner or later the cricket was bound to resume, and later he did, revealing his hideaway behind the refrigerator.

In the process of trying to remove him from that awkward recess I inadvertently dispatched him to Elysian fields where presumably he joined choruses of Elysian crickets. My neighbor thanked me profusely. I was sad. The episode did remind me, however, that the season of crickets and grasshoppers was at hand.

To me there is a sadness in the singing of the orthopterans. Birds sing of the coming of summer, of a beginning. The orthopterans sing of the end of summer and the coming of fall. Birds sing of many summers; the orthopterans of one. For them there will be no others. There seems to be a haste, a frenzy, and a persistence in their singing, almost as though they knew that time was short.

Most people are unaware of insect singers unless there happens to be a cricket in the house or a garden full of katydids arguing the night away. So I listened, and then I heard a tremendous low murmuring, tinkling, and trilling of literally thousands of ground crickets. These small black and brown crickets swarm in the fields and pastures, and our lawn was no less populated. Listening attentively now that I was atuned, I distinguished the buzzing chirps of the striped ground cricket and the slow trill of Allard's ground cricket, two species distinguishable more by their songs than by their appearance. As I crossed the lawn, the chorus would stop as dozens of crickets scrambled to safety. Elsewhere the noise continued, sibilant in its unison.

Walking to the place where the forest bordered the lawn I could pick out the tintinnabulations of the woodland shade-loving tinkling ground cricket. Had I ventured down to the small sphagnum bog near the brook I would have heard the soft and delicate buzzing of marsh ground crickets because this was their season too.

Near the garden, where the grass along the bordering stones was always moist, the shrill stuttering of the Carolina ground cricket signaled its domain. From across the road in the taller grass of the hayfield the buzzing of the meadow grasshoppers, the coneheads, and the meadow katydids, suited perfectly the swishing and rustling of the tall, pollen-laden stems.

Then as my ear began to sort out the many choruses, I heard the clear solos of the field crickets, the chirp, chirp, of each singer ringing from his own personal burrow or crevice. Several sang from various retreats in the low stone wall bordering the garden. There was one in the steps to the side porch. Another called from a crack in the granite steps to the front porch.

These large, handsome, black crickets are the ones familiar to so many people and abhorred when discovered indoors. They are relatives of the ones that are used in cricket fights in China and for fish bait in our country.

In southern New England there was confusion about these crickets for many years. They are the first insectan singers to appear in the spring and among the last to disappear in the fall, but in the middle of the summer they are not to be found. We now know that there are two distinct species, identical in appearance and song, but distinct in life history. The northern spring field cricket spends the winter as a partially grown cricket, a nymph, ready to reach maturity in the spring. When the pussywillows and popular catkins have turned to fluff and cast their silk-cotton to the breeze, the spring field crickets have already begun their serenades. By midsummer they have lived, mated, and died, their places taken by the northern fall field crickets. These, having over-wintered as eggs, need half the summer to reach maturity. Around the bungalow we had the fall crickets exclusively.

A very friendly male lived in the steps to the front porch. These steps were sadly in need of pointing. The old mortar had rotted to such a condition that small dandelions, devil's paintbrushes, plantains, and grasses grew very well in the crevices. It almost became necessary to weed the steps. I had put off the pointing because it actually meant resetting some of the blocks. At the moment I was glad of my negligence because delay had provided several neat retreats for crickets. Psychologically there was room for only one because the males are rigidly territorial and space themselves out rather widely.

At first this particular male was wary; any movement close to his retreat silenced him. On the other hand, it was possible to approach more closely to him than to crickets in the field because the solidity of the granite precluded vibrations being set up by our footsteps. Thus, unwarned, he failed to detect our presence until alerted by sight of us or our shadows.

As time passed, however, he became less shy. Eventually we could even sit on the step by the entrance to his cave and watch as he serenaded from his doorway. He was, of course, serenading a female whom he had never met. In fact he was serenading any and all females. In a reversal of the usual social rules the females wandered about in the grass while the males sat at home. If the cricket in the steps was lucky, some female, indeed several females, would be enticed by his song and come seeking.

I could not honestly claim that this cricket in the steps was a resident of the bungalow, but then I did not have to. One foggy night we were singularly honored—or cursed—depending on one's point of view. The family, Gulliver included, was gathered around the fireplace (it was too warm for a fire despite the fog) and engaged in one pursuit or another, reading, writing, or tinkering, when unexpectedly from a chink in the rocks of the fireplace there rang out a bold chirp, chirp, chirp, chirp. We actually had a cricket on the hearth.

From time immemorial a cricket on the hearth has been a sign of good luck. No lines can convey the sentiment of cheer in the sombre lights and shadows of the bungalow better than those of Milton:

> Where glowing embers through the room
> Teach light to counterfeit a gloom,
> Far from all resort of mirth,
> Save the cricket on the hearth.

Milton was referring of course to the European house cricket, since then introduced into our country; ours was kin to the native in the stone steps.

Each of us expressed pleasure at the presence of the new arrival. We did not share my neighbor's sentiments, nor were we concerned about the cricket chewing our rugs or vandalizing our belongings. Later that night when we had all retired, the last sound we heard was the cheery chirping. We dozed off, secure in the knowledge that we had acquired a new watchdog—with apologies to Gulliver.

From that night onward our cricket was tireless in his chirping serenade. Receiving no encouragement he nonetheless sang all afternoon and long into the night. If for some reason he interrupted his lay, we became immediately aware of the silence. To us the chirping never became wearisome. It occurred to me, however, that we could ask him for a different song if we wished. His repertoire, though not extensive, was varied. His choice was determined by circumstances. All I had to do was change the circumstances.

Leaving him to his singing I went to the garden where I listened attentively for a few minutes. Having learned what I wished to know, I went directly to a particular rock. Turning it over quickly I found what I expected, a male interrupted in midsong. Before he could escape I coaxed him into a small pickle jar. That evening while "our" cricket was singing on the doorstep of his crevice in the hearth, I removed the lid of the jar containing the captive from the garden and laid the jar gently about eight inches in front of the singer. He in his turn hesitated only momentarily in his song. The stage thus being set I stretched out in front of the hearth to watch. There is no reason why one should not be comfortable while observing nature.

It took about a minute for the captive to discover that his prison door was open. Undoubtedly he also heard the other male calling. Rather cautiously he crawled from the jar and advanced toward the singer. That cricket stepped forward from the crevice. His calling stopped. After many hesitations and pauses the two crickets came within antennal reach of each other. For a moment they stood face to face as though trying to stare down one another. Next they began to feint with their antennae. Now like two epee duelists they engaged in fierce antennal lashing, neither giving ground. The frenzy increased as first one then the other leapt to the attack. Our male reared up and lunged forward, accompanying his actions with a fighting song. Whereas the calling song had sounded somewhat musical, clear, precise, and regular in beat, the battle song consisted of long, sharp, less melodious chirps. They actually sounded more agitated.

The intruding male responded in kind, and the violence of the encounter increased further. The two crickets began to bite each other, then to rear up and spar with their front legs. For two or three minutes they bit, pushed, and butted. Now both were chirping battle songs.

In a few seconds, the fight was over. Our cricket, stronger or emboldened by the psychological advantage of fighting on his own territory, suddenly flipped the intruder into the air. He landed on his back, recovered quickly but, vanquished, began to retreat. At this point I recaptured him for return to his home in the garden. As far as I was concerned, he had served my purpose. He had induced our male to change songs. Our cricket continued his aggressive chirping, but by the time I returned he had reentered the crevice at the entrance of which he had resumed his customary calling as though nothing untoward had happened.

I understood now the fascination that cricket fights hold for the Chinese. A person knowledgeable in the way of crickets could select from the best population and from among the best fighters as ascertained by preliminary bouts. According to Chinese lore the best fighters in the Hong Kong area were human-head crickets. They derived their sobriquet from the circumstances of their living, their dwelling places being human skulls in overgrown and crowded graveyards. With a stable of good fighters a person with gambling instincts could do quite well for himself. This was the motivation underlying cricket fighting.

Having succeeded in eliciting from our cricket a performance of battle tunes, I tried a few days later to evoke a different song. For this I had to capture a female. Finding a female turned out to be a difficult task because females lived neither in settled homes nor in circumscribed territories. Neither did they reveal their whereabouts by singing. Finally, to save myself the time of much aimless searching, I left a few apple cores in places that I imagined might be frequented by females. This stratagem succeeded; so, supplied with a vigorous, robust female I repeated the tactics I had employed with the intruder male.

At the initial face-to-face encounter the male led off with antennal fencing indistinguishable from that which he had performed with the intruder male. The female, however, immediately became docile and unresponsive. To the male, immobility and docility apparently signified female because he promptly switched from calling to his courtship song, a series of soft rhythmical chirps punctuated with clicks. We attended to this song for a while, but it was not only soft, it was ephemeral. More important matters were about to begin. Hardheartedly I retrieved the female for return to the lawn.

The cricket remained on our hearth for the balance of the summer.

When, toward the end, the nights grew colder and a fire was built, he merely exchanged his crevice for one out of range of the heat. From there he would sing as lustily as ever. On our final day, with the bungalow boarded up for the winter and we ready to lock the door, the last sound from inside was the undaunted chirp of a supreme optimist. Under normal circumstances in days of yore the housebound cricket would have found himself in a winterized dwelling where the warmth of the fire would keep him warm long after snow had fallen. The cricket on our hearth would survive through Indian summer if one of the wood mice had not found him before then. The bungalow was no protection against the cold of November.

Perhaps fortune would favor us again the next summer, and we would have another cricket on the hearth.

15 ∘ THE ROOT CELLAR

F̲o r several decades after down-easters stopped shipping ice to South America and Africa in wooden schooners, they continued to cut pond ice for domestic use. Once or twice a week, depending on the weather, the iceman would come in his horse-drawn wagon, cut the desired cake of ice, trim it as it hung from the scales, brush off the sawdust remaining from its being stored in the icehouse, and carry the piece to the icebox. The most perishable foods were kept in the icebox. Important adjuncts to iceboxes were springhouses and root cellars. The bungalow had been built with a root cellar accessible from the outside by way of a bulkhead.

Ordinarily a root cellar served two purposes. In the winter it was a place of storage for root crops, beets, turnips, and carrots, as well as squash and apples, a place intermediate in temperature between the freezing of outdoors and the heat of indoors. Generally it was dry, as root cellars should be. In the summertime it remained cool enough even on the hottest days to preserve butter and milk as well as other perishables.

When iceboxes were replaced by refrigerators, root cellars fell into

disuse. We continued to use ours as a place to store clams and lob-
sters overnight. It was also an admirable cave in which to store
homebrew.

Since the bungalow was not winterized, its root cellar did not remain dry. It often became as damp as a springhouse. In early spring
and during prolonged rainy periods it took on the appearance of a
cistern as the ground-water level rose. Because of these characteristics it became home for animals that found its damp, cavelike features reminiscent of their natural habitats.

As had happened other times in our relations with inmates of the
bungalow, we had not actively sought for animals in the root cellar.
The discovery was accidental. Earlier in the day the boys and I had
gone clamming. Clams were scarce this year either because of some
normal population cycle or because of over-fishing by commercial
diggers. As a consequence, it was only after considerable labor that
the three of us managed to fill a hod. Upon our return, as was our
custom, we transferred the clams to two ten-quart pails which we
then filled with seawater. These pails were placed in the root cellar.
My grandfather used to add cornmeal to freshly dug clams in order,
he said, "to clean them out." We found that an overnight rest in
plain seawater got rid of most of the sand just as well. That evening,
however, we decided to have a few steamers for supper regardless
of the sand; so Jehan went out to select some.

"Hey!" he called from the depths of the root cellar. "There's
something funny here."

The rest of us were busying ourselves in the kitchen and did not
particularly care to interrupt our activities on the basis of such an un-
informative exclamation.

"What do you mean 'funny'?" I called back.

"Well, it's part way in a hole in the rocks so I can't really see it, but
I think it's a salamander."

This *was* interesting news, so I stepped out to investigate.

"It's in there," Jehan said, pointing to a gap in the rocks where the
top step joined the wall.

Our root cellar was of dry-wall construction, that is, unmortared.
By and large the construction was in splendid condition; the rocks
still tightly in place, but here and there age or frost had loosened

a stone. At the place pointed out by Jehan a loose stone left a rather large crevice. Gingerly I pulled out that stone, making a mental note at the time to mortar it in place later. That good intention was not to be executed soon, however, because within the cavity lay a salamander obviously quite at home. Interested though we were, this was not the time to pursue natural history; clams waited to be steamed, butter to be melted, supper to be eaten. The salamander would wait.

The following morning the boys and I went to inspect our newly discovered resident. I removed the stone and saw that he was still there. Reaching in carefully I removed this gentle creature, and brought him into the light for clearer observation. He was beautiful, about six inches long, sleek, deep blue-black, and decorated with large yellow spots. During the inspection he made no effort to escape.

How maligned the salamanders have always been! These harmless amphibians, lovers of coolness and darkness, seekers of moisture, creatures of water, were so misunderstood in time past that they were associated with fire. It was believed that they could actually live in fire. From this belief, so irrational considering the moisture of the animal's skin, was derived the application of the word "salamander" to all sorts of objects dealing with fire. Not many decades ago one could purchase in the general store in East Bluehill, a salamander, that is, a metal plate that was heated and held over food to brown the top. By no stretch of the imagination could our salamander be seen as fire resistant, fierce, or malevolent. He was the most secretive of the bungalow's inhabitants. Had we not stumbled upon him accidentally, it is doubtful that we would ever have learned of his existence. Now, however, having discovered him, we became interested in his comings and goings. Occasionally I had seen one of his kind under a moist rotten log, but I had never paid particular attention to the species.

The life of a salamander must be one of excruciating boredom. Adults of the spotted salamander spend five or six months of the winter hibernating beneath rock ledges or in debris or soil at the bases of trees. In stark contrast to their mythical attributes they are able to withstand freezing cold. In very early spring, often while there is still snow on the ground, they emerge from their winter quarters. Traveling at night they strike out across the snow for the nearest pond. There are reports of their crossing roads in large numbers only to be thwarted by snowbanks thrown up by plows. Failing in their attempts to surmount the bank they travel parallel to it, seeking, as it were, a pass over the mountain to the pond beyond. Sometimes the pond is a vernal body of water originating from melting snow and spring rains, a pond destined to disappear in the heat of midsummer.

Salamanders are known to hibernate as far as one hundred and fifty feet from the nearest water although more conservative individuals may not stray farther away than fifteen feet. In either case, about the time that the frogs, spring peepers, and toads are beginning their spring choruses the salamanders are also congregating to breed. Their courtship, however, is a silent one. Their vocalizations are limited to occasional unmusical bleats, burps, grunts, or wheezes. Some sense other than the perception of siren songs draws them together.

Their eggs, resembling small helpings of sago pudding, are fastened to vegetation sticks or branches bent into the water by burdens of snow. Sometimes tragedy overtakes eggs laid on branches the tips of which are frozen in ice. If the ice melts before the eggs hatch, the branch, released from the grip of the ice, springs into the air. There the eggs rapidly dry and shrivel. Eggs in the water develop into larvae resembling elongated tadpoles with four legs and bushy gills. In the meantime the parents have lost all interest in eggs and larvae. Although they may remain in the pond for two or three weeks, they then return to land, each going its separate way.

From this time on the salamander resides in a summer home where it remains in solitude for the next five or six months, idling away the summer days and emerging only at night to feed. Obviously the root cellar was serving as a summer home for one spotted salamander. But could it be more than a summer home? Certain unique characteristics of the root cellar offered other potential attractions. For example, the cellar flooded annually with the melting snows. Occasionally it remained flooded through June. Could salamanders breed there? It was an interesting possibility but one we could not verify without a trip to the bungalow during the spring. A second possibility, also difficult to confirm, was that the root cellar served as winter as well as summer quarters. For an animal as hardy as the spotted salamander, hibernation in the cellar was feasible, but during the winter deep snows weighed down the bulkhead so that we would be able to enter the cellar only at the expense of much digging.

Considering the root cellar in summer as a unique ecological niche, dark, cool, moist, and, most important, secure from skunks, raccoons, and other predators that might relish plump, juicy salamanders, I wondered whether our salamander spent his waking as well as his sleeping hours there and whether he also did all his foraging in the same place. We knew that he ate insects. Beetles, grubs, other larvae of all sorts, and probably crickets, were gourmet items from his point of view. How many of these, if any, did the root cellar supply?

Prompted by these musings I began a careful search. In the process I discovered that our artificial cave was home for a number of animals found nowhere else in the house. Most conspicuous were the cave orb spinners. I counted twenty-five, strikingly similar in appearance to the cobweb spiders of the comb-footed family. Every corner of the cellar was webbed with their inclined orbs. Fluffy white bundles of eggs hung by threads from the walls. There in the perpetual darkness these spiders lived out their lives, seeing light only when we opened the bulkhead to the cellar.

It seemed most unlikely that the salamander feasted on these spiders. For one thing, they lived in high corners and on the ceiling among their many webs while the salamander spent his life on the ground. Encounters were unlikely.

My continued search uncovered more probable candidates. Also living among the rock crevices were several camel crickets. Strange, wingless, humpbacked orthopterans, these mute, secretive crickets are seldom seen except by collectors. They are excellent jumpers and could be very agile when pursued, but usually they depended on their mottled brown ground color to protect them. Oftentimes, however, the waving of their extraordinarily long, slender antennae betrayed their camouflage. We had them in our woodshed, but these were the first I had observed in the house. Like so many cave dwellers they were shy and harmless, feeding on organic debris, cleaning up the crumbs and remains of others. They would suit the appetite of the salamander perfectly. Their only safety lay in concealment.

Several times at night I checked to see if the salamander was at home. Sometimes he was; sometimes he was not. I also checked the camel crickets periodically. They were always at home. Clearly the salamander was not feasting on them. He must, therefore, have spent some of his nights foraging for food. His continued presence during that summer, and the succeeding one as well, indicated that he knew where his home was and how to find it when returning from trips.

These salamanders in general often return to the same pond from their winter quarters year after year. Whether this indicates a homing ability or merely a sense that guides them to the nearest body of water is not clear. There is much to be learned about them, but, except during their spring and fall migrations, they are seldom observed.

The salamander in the root cellar spent at least two summers there. What happened to him after that is only conjecture. His species is known to live for more than three years, perhaps as long as ten. Likely as not he was surprised one night by a raccoon or some other large predator.

The other denizens of the root cellar had shorter life spans, but few of them ever ventured afield; consequently, each succeeding generation inherited the cellar. It is interesting to muse about this inheritance. One of the unique characteristics of human beings is that of bequeathing material things. Here, in a sense, are insects and spiders bequeathing property, that is, territory. True, there are no formalities; it is merely a question of squatters' rights. Yet, lineal descendants by virtue of having been born where their parents had lived, occupied the same cave, or cellar, or corner—unless, of course, they were driven off by aggressors. But is not the same as true for us?

The root cellar was just one specialized ecological niche offered by the bungalow; the exposed sun-baked stone of the chimney, the soft dry wood of the cedar posts, the windless predator-free protection of the interior, the shelter of the eaves, the unique construction of the porches. The bungalow, as I had surmised that morning on the roof, was inhabited from chimney to cellar.

16 ∘ UNDER THE SIGN
OF ORION

WE arrived at one o'clock in the morning. Because of a five-foot snow cover we were obliged to tote supplies in from the road on a toboggan. It was cold, just three degrees above zero. The stars glinted like ice crystals from the frigid blackness of space. It was a Christmas sky, a sky with eight stars of the first magnitude. Overhead the hunter Orion and his two dogs were on their annual winter passage through the heavens.

We stumbled past the silent bungalow, huddled in its own mantle of snow, to the cabin in the woods that served as a guest house during the summer. On these winter trips we always stayed in the cabin. Neither cabin nor bungalow was winterized, but the cabin had the marginal advantage of a small pot-bellied stove. During this mid-winter stay the boys and I would rely for all our water on water brought in from the town spring. Candles provided the principal light, but during meal times we supplemented them with the luxury of a twelve-volt bulb run from a storage battery, courtesy of Jehan. Meals could be cooked on the pot-bellied stove which also brought the room temperature up to a point where we no longer saw our

breath. When the fire went out, the temperature rapidly returned to the outside low. Needless to say, on the night of our arrival we wasted no time admiring the heavens. We climbed immediately into our sleeping bags. Sleep escaped us only as long as was necessary for the bags to warm to a tolerable temperature. During that brief period I breathed deeply of the freshness of the winter air and listened to the night. And I heard nothing. If anything stirred, it moved silently. As far as our senses could tell we were totally and utterly alone in the winter darkness. And so sleep came.

All the next day it snowed, so we divided the time between getting organized and exploring. The only relief from whiteness in field and forest was the dark shape of the bungalow, the black trunks and branches of trees, the dark, ever-so-dark green of the conifers, deep almost funereal hues accentuated by the mantle of snow that was already bowing down the branches. Only to seaward were there hints of other colors. Rocks on the shoreline lacked the hues of summer, but contrasts of blacks and browns, draped at the low-tide mark with yellow-brown rockweed, broke the somberness of black and white. The sea itself, in contrast to the snow, was the grayest of grays. Most noticeable of all was the apparent absence of life.

The bungalow also seemed utterly devoid of life. I stood looking around on the back porch. On the beams and wires were the nests of robins. Sheltered as they were from the elements their mud walls had not begun to erode nor their grassy foundations to unravel as was the fate of nests in trees. These were as the fledglings had left them and they appeared not so much abandoned as waiting. About them was an air of expectancy. The robins themselves were worrying about other matters in warmer climates. Not until longer southern days and endocrine urges directed them northward would they remember these haunts.

I looked at the cedar posts on which there was not a single sign of life. There were the holes of the community of tiny wasps, looking the same as in summer except that they now lacked the entrance collars of sawdust. Within the tiny galleries, the mummylike pupae rested in a mysterious suspended animation, neither lifeless nor lively. Within the cells of their brains-to-be existed the potential behavior of their generation as well as all future generations. Similarly future wasps in the caves in the chimney were essence rather than being. The snowflakes fell upon me as I stepped off the porch, but nothing buzzed around my head.

I waded through the drifts along the side of the bungalow and peered up at the eaves. Like the nest of the robins the gray paper Japanese lanterns and chandeliers of the bald-faced hornet and Polistes hung unaltered from the summertime, as though ready for renewed occupancy during the summer to come; but there would be no return of the old tenants nor replacement by new. Those that built the nest were dead. Only the queens born there lived, and they, only if they had escaped predators; if so, they were hanging somewhere in a torpor so deep that it mimicked death. They had hooked their claws in some rough surface so securely that no muscular effort was required. If perchance they died before awakening, their bodies would be found still hanging, no less lifelike than before. The lucky ones, some of them hibernating somewhere in the bungalow, would stir themselves every time, however briefly, the sun warmed the air until eventually summer arrived to stay. Then they would begin their own paper masterpieces.

There was life under the eaves if one knew what to seek. Not concealed in holes or crevices but most often in open corners there were numerous fuzzy, light brown bundles, meshed in webbing stuck between every pair of rafters the entire length of the eaves. These were all that remained of the impressive orb spinners. Gone were the exquisite embroideries of the summer, too fragile to have withstood even the gentler breezes that the eaves transformed from the greater winds of the open spaces. Gone were the creators of those webs. Of them the only remains were a few shed skins, wizened mockeries of the robust spiders of yestertimes. The reminders of promise were the brown bundles of eggs. Safe from most predators, protected from the harsher elements, the eggs with the advent of warm weather would yield hundreds of tiny spiders.

When I got around to the front porch, the complete force of the storm hit me as it drove in from the ocean. Even the open front porch afforded some protection. High up on the beams and rafters the protection was nearly complete. The nests of the swallows and the juncos, like those of the robins, seemed to await occupants. The juncos had moved south, but not so far as the robins. They were still in snow country but where winters were less severe. Some probably foraged on seeds spilled from feeding stations in back yards. The swallows were far, far, away in the land of mangos, papayas, and torrid tropical days.

The snow was falling faster, the temperature dropping. This was as good a time as any to go into the bungalow to check on the shape of things, leaks, weather damage, animal vandalism, or signs of entry by humans.

My first impression was of quiet, darkness, the cold of the grave. The quiet came as a striking contrast to the gales outside. The wind did not howl because the snow-laden trees could not respond. The snow muffled all the other sounds, and it did not beat on the roof as does the rain. The cold inside struck me as more penetrating, more deathly than outside because there was no movement of air, not the slightest. It was an implacable cold. Even the far-off caves to which the little brown bats had migrated held enough residual heat in their inner recesses so that the bats were able to maintain a measurable level of metabolism even though their fur might be covered with hoar frost. In the winter bungalow they would have frozen. It would have been a trap rather than a refuge.

The carpenter ants stood frozen in their galleries, each one mobilized in the act of whatever she had been doing when the first freeze crept over the colony. Often when splitting logs in wintertime I had exposed such colonies. In contrast to their summer frenzies they reminded me of a motion picture that had suddenly been stopped. If I breathed upon them, they would move briefly in slow motion only to stop once more in some new, half-executed posture. It could have been a picture of the last movements of ants in fossil amber, trapped in the sticky mass, suffocated in one last gesture, and fossilized for all time. I did not have to see into the beams of the bungalow to picture the scene there. It was a scene repeated in nearly every dead tree in the forest.

There were telltale signs of wood mice, but none of squirrels. I was sure that numbers of warm nests had been constructed in inconspicuous corners. A quick examination showed that the plastic that encased the overstuffed furniture and pillows had no holes, so I felt reasonably sure that the mice had found quarters elsewhere.

The probability of there being other life in the bungalow was strengthened later that night in the cabin. The best efforts of the stove sufficed only to bring the temperature of that corner of the cabin up to forty degrees Fahrenheit. We roasted in front and froze behind, but the heat in the corner had awakened a house spider. Unlike the orb spinners some house spiders live more than a single year. One of these, duped into believing that summer had arrived, was laying a thread along my pant leg.

The next morning, while the sun still sheltered behind Cadillac Mountain, we were awakened by the sound of many ducks. Since it was no colder out than in, we added mittens and caps to what we already wore and stepped out into a glorious morning. Now all about us there were signs of life. The flat calm bay just beginning to turn softly golden was a patchwork of ice pans and open water. Everywhere in the ice-free spaces rafts of sea ducks talked to one another in a language difficult to describe. Their calls were not the quacks of the river ducks; they were high-pitched gabbles. With the field glasses we saw scoters, eiders, old squaws, goldeneyes, and black ducks. Here and there a dovekie changed from one patch of open water to another. Herring gulls had begun their morning patrols. On one of the headlands the crows began an argument.

Turning from the shore we saw evidence of life everywhere. From the bungalow numerous dainty tracks confirmed my earlier conclusion that we had indeed failed to render the place mouse-proof. A snowshoe hare had bounded through the garden. Perhaps he was even then watching us from an open spot, his winter coat of white blending insensibly into the snow. On the road a fox had trotted in the general direction of the brook before detouring into the woods. Perhaps he had his thoughts on the spruce grouse whose prim tracks crossed the road at that precise spot.

In the trees behind the woodshed a small flock of chickadees, including a rare Hudsonian chickadee, and some red-breasted nuthatches started small avalanches of snow in the laden branches as they sought breakfast. Then we saw two red squirrels. Perhaps they had already seen us; but when we stopped to look, they began to scold. The temperature still hovered around zero; nevertheless, the world had come alive. Even on the surface of the snow itself a fine black peppering, like the metallic dust from a grinding wheel, revealed itself as thousands of snow fleas—not fleas at all but springtails, those most ancient of insects. Picking up a fistful of snow and breathing upon it I brought these motes briefly to life. Later when the sun rose higher they would spring over the snow in quest of who knows what.

Had we explored further, we would have discovered more animals abroad that I could catalogue here. It was precisely these observations that enabled me to delineate the ecology of the bungalow.

Our bungalow was a summer house indeed. With the exception of the wood mice, the squirrels, the lichens, and, possibly, the salamanders, all the species that inhabited it were summer animals. When summer ended, they migrated, hibernated, within or without, or died, leaving as legacy the germs of the next generation. During the summer the bungalow offered shelter and protection.

The contrast between those who sojourned in the bungalow and the year-round residents was striking. For the latter the building also could provide shelter from the elements and protection from predators. Understandably it had a powerful attraction for wood mice and squirrels. But it was a bare larder. Moreover, it lacked the proper architectural context for birds of the trees, fields, forests, for the grouse, chickadees, nuthatches, kinglets, woodpeckers, and others. It had nothing to offer the fox, the mink, the weasel. Those adapted to thrive in Maine winters did not need a bungalow.

Later that afternoon as we made preparations to leave, storm clouds again rolled in from the west. Soon snow began to fall. The bungalow appeared forlorn, deserted, barren. Of the creatures that enlivened it in the summer only we could bring back the memory of light and color and gaiety of year after year, and only we could see ahead, however imperfectly, to another summer.

As I turned away the last thing I saw before a heavy shower of snow flakes obliterated all form and substance was a ghostly snowy owl gliding across the field toward the shelter of the forest.

THE RESIDENTS OF
THE BUNGALOW

In the chimney and rocking chair
Mason wasps (Eumenidae)   *Symmorphus sp.*

In the roof
Carpenter ants (Formicidae)   *Camponotus pennsylvanicus*

In the cedar posts
Large wasps (Pemphredoninae)   *Passaloecus cuspidatus*
Small wasps (Pemphredoninae)   *Spilomena ampliceps*

In the door trim
Mud dauber (Crabronidae)   *Rhopalum pedicellatum*
Northern flicker   *Colaptes auratus luteus*

In the living room
Little brown bat   *Myotis lucifugus*
Wood mice   *Peromyscus maniculatus*
Red squirrels   *Tamiasciurus hudsonicus*
Cobweb spiders   *Theridion sp.*
                                           Achaearania tepidariorum
                                           Steatoda borealis

The garden spider   *Agelenopsis sp.*
European house spider   *Tagenaria sp.*

On the hearth
Field cricket   *Acheta pennsylvanicus*

On the back porch
Eastern robins   *Turdus migratorius migratorius*
Bumblebees   *Bombus vagans*

On the front porch
Cliff swallows   *Petrochelidon albifrons albifrons*
Barn swallows   *Hirundo erythrogaster*
Slate-colored juncos   *Junco hyemalis hyemalis*

Under the eaves
Bald-faced hornets (Vespidae)   *Dolichovespula maculata*
Slender paper wasps (Vespidae) *Polistes sp.*
Orb spinners   *Araneus trifolium*

Above the kitchen window
Phoebe   *Sayornis phoebe*

In the root cellar
Spotted salamander   *Ambystoma maculatum*
Cave orb spinners   *Meta menardii*
Camel crickets   *Ceuthophilus maculatus*

Everywhere
Gulliver   *Canis familiaris*
Paul, Jehan, Lois, Vincent   *Homo sapiens*

SUGGESTED READINGS

Alexander, R. D.; Pace, A. E.; and Otte, D. 1972. The singing insects of Michigan. *Mich. Ent. Soc.* 5(2):33–69. Up-to-date keys and names. Also keys to songs.

Allen, G. M. 1962. *Bats.* New York: Dover. A classic and readable account of the natural history of bats. Contains much descriptive material not found in more up-to-date treatments.

Banfield, A. W. F. 1974. *The Mammals of Canada.* Toronto: University of Toronto Press. A beautifully illustrated and informative book describing appearance, habits, habitat, reproduction, and distribution. Relevant to all North America.

Bent, A. C. 1963. *Life Histories of North American Flycatchers, Larks, Swallows, and Their Allies.* New York: Dover. A most detailed treatment of the birds mentioned in the title.

――――. 1964a. *Life Histories of North American Thrushes, Kinglets, and Their Allies.* New York: Dover. The standard reference.

――――. 1964b. *Life Histories of North American Woodpeckers.* New York: Dover.

――――. 1968. *Life Histories of North American Cardinals, Grosbeaks, Buntings, Towhees, Finches, Sparrows, and Their Allies.* U.S. Nat. Hist. Mus. Bull. no. 237. Washington, D.C.: Smithsonian Inst. Press.

Bishop, S. C. 1947. *Handbook of Salamanders.* Ithaca, N.Y.: Comstock Pub. Co. Descriptions of the habitats, distributions, characteristics, and life histories of American salamanders. A useful guide.

Blatchley, W. S. 1920. *Orthoptera of Northeastern America.* Indianapolis: Nature Pub.

Co. Still one of the richest sources of locust, grasshopper, and cricket natural history. For current classification and naming consult Alexander et al.

Bristowe, W. S. 1958. *The World of Spiders*. London: Collins. Although dealing mostly with British species, this is a rich source of information about spiders in general.

Evans, H. E., and Eberhard, M. J. W. 1970. *The Wasps*. Ann Arbor: Univ. Mich. Press. An excellent, enjoyable account. Contains much information on mason, digger, paper, and mud-daubing species.

Gertsch, W. S. 1979. *American Spiders*. 2d ed. New York: Van Nostrand and Reinhold. The American equivalent of Bristowe. An excellent introduction to spiders.

Griffin, D. R. 1958. *Listening in the Dark*. New Haven: Yale Univ. Press. A modern treatment of bats with emphasis on their hearing and hunting.

Headstrom, R. 1973. *Spiders of the United States*. New York: A. S. Barnes and Co. A concise handbook for identifying spiders.

Kaston, B. J. 1978. *How to Know the Spiders*. 3d ed. Dubuque, Iowa: W. C. Brown, Co. An excellent introduction with illustrated keys for identification.

King, J. A., ed. 1968. *Biology of Peromyscus* (Rodentia). Special Pub. no. 2, Amer. Soc. Mammalogists. A collection of detailed reviews by many authors covering all aspects of the biology of wood mice, deer mice, and their relatives.

Landy, M. J. 1967. A study of the life histories of two sympatric species of amblystomatid salamanders. Ph.D. dissertation, Univ. Mass. A nontechnical description of the biology and life histories of the spotted salamander and a near relative.

Levi, H. W., and Levi, L. R. 1968. *A Guide to Spiders and Their Kin*. New York: Golden Press. A profusely illustrated field guide for beginners. A fine introduction.

Michener, C. D., and Michener, M. H. 1951. *American Social Insects*. New York: D. Van Nostrand. An authoritative and entertaining account of bees, ants, wasps, and termites.

Pierce, G. W. 1948. *The Songs of Insects*. Cambridge: Harvard Univ. Press. An early descriptive and analytic study of singing insects of New England written for nonspecialists.

Plath, O. E. 1934. *Bumblebees and Their Ways*. New York: Macmillan. Still the most personal, entertaining, and authoritative account of bumblebees. Gives details for identification.

Rau, P., and Rau, N. 1970. *Wasp Studies Afield*. New York: Dover. A charming classic dealing with the authors' personal and original field studies.

Shorten, M. 1954. *Squirrels*. London: Collins. The biology of squirrels with emphasis on European species. Written for naturalists.

Walker, E. P. 1975. *Mammals of the World*. 3d ed. Baltimore: The Johns Hopkins Univ. Press. A complete and detailed reference to mammals.

Welty, J. C. 1962. *The Life of the Birds*. Philadelphia: W. B. Saunders. A comprehensive, entertaining account of birds. Contains a wealth of information not available elsewhere.

Wheeler, W. M. 1910. *Ants*. New York: Columbia Univ. Press. A classic. Much of the exhaustive information is still valid despite the age of the book.

Yalden, D. W., and Morris, P. A. 1975. *The Lives of Bats*. New York: Quadrangle, N.Y. Times Book Co. A concise, easily read introduction to bats.

Library of Congress Cataloging in Publication Data
Dethier, V. G. (Vincent Gaston), 1915–
The ecology of a summer house.
Bibliography: p.
1. Ecology—Maine—Miscellanea. 2. Summer—Maine—
Miscellanea.   I. Title.
QH105.M2D48   1984   574.9741   83-18007
ISBN 0-87023-421-8
ISBN 0-87023-422-6 (pbk.)